Analysis of Catastrophes and Their Public Health Consequences

Paolo F. Ricci

Analysis of Catastrophes and Their Public Health Consequences

Descriptions, Predictions, and Aggregation of Expert Judgment Supporting Science Policy

 Springer

Paolo F. Ricci
University of Bologna, Ravenna Campus
Ravenna, Italy

ISBN 978-3-030-48068-4   ISBN 978-3-030-48066-0   (eBook)
https://doi.org/10.1007/978-3-030-48066-0

This Springer imprint is published by the registered company Springer Nature Switzerland AG
The registered company address is: Gewerbestrasse 11, 6330 Cham, Switzerland

To Andrea, my wife: her intellectual and emotional support is perennial.

Preface

In this Book we deal with the analysis and evaluation of *prospective* (probabilistic and future) catastrophic incidents generally categorized as either natural or man-made (*human-made*). These range from earthquakes to economic recessions and, depending on the size of their consequences and frequency, can be routine, non-routine, and extreme: *dragon-kings* and *Black Swans*. These events' key characteristics may be briefly summarized by coupling the size of their consequences with their probabilities of occurring. For policy decisions, informed by the assessment of available choices within the overall context of public health science-policy, this summary representation may be necessary but is insufficient. The reason is that diverse causes, mechanisms, and consequences characterize each catastrophe: even distributions do not explicitly indicate the mechanisms relating consequences to probabilities. Our work allows readers to more completely understand, without imposing burdensome analytical demands, the assessment of precautionary choices, including the option not to choose and to wait for additional knowledge. Working in reverse, starting with a proposed choice, our work allows assessing an assumed cause and effect model before the catastrophic incident being modeled occurs. By reducing complications, we provide a simple but comprehensive exposition of the science-policy issues that affect assessing prospective catastrophic incidents. The issues we discuss range from legal-regulatory requirements to determining the impact of infectious diseases, from floods and tsunamis to cancer incidence from environmental exposures. Critically for science policy, the scientific assessments also require expert advice – from multiple and diverse disciplines – regarding the use of data, models, and the results. The aggregation of expert opinions, through either ranking or voting, identifies the collectively preferred choice and completes our work. An analogy describes it. The legal basis for making decisions on behalf of society threatened by prospective catastrophic accidents is the roof. It rests on a truss with struts, beams, posts, and other structural members: inputs and outputs that connect cause to effect. Its pillars are the sciences; the foundation is the prospective reality being assessed. Uncertainty analysis and mechanistic causation, developed through theoretical and empirical reasoning about events, risk factors, interactions, and feedbacks, define the skeleton of the architecture of the eventual policy structure.

Although the cost of this simplification is a thin discussion, it opens the door to understanding how to identify optimal or preferable precautionary choices through formal descriptions of a prospective reality through predictions. Our work includes criteria that yield a preferential ordering between precautionary choices using formal criteria that account for the continual evolution of scientific and technological knowledge and information, K&I. This evolution informs policy makers, stakeholders, and the public while accounting for second best and other solutions.

The guiding policy objective is maximally to protect society while spending the least amount of scarce resources. Protection often requires large and certain expenditures directed against catastrophic incidents that may recur infrequently and unexpectedly. When one of those catastrophes happens, it can dislocate societal functioning and result in extreme damage such as hundreds of thousands of prompt deaths. However, before the catastrophe occurs, these consequences are characterized by very low probability of occurring. This raises the issue of determining the appropriate and achievable level of protection that society should adopt. For routine catastrophic events, the issue has been resolved through best engineering practices, advanced materials and designs, building codes, international codes of practice, and many other procedural and substantive methods. But, protecting against prospective rare or extremely rare events is affected by physical and budgetary limits regarding what can be realistically be done – rather than what should be done. For instance, having to spend scarce resources now to protect against an event that may *on average* recur once in several hundred years or more creates a difficult societal dilemma. This dilemma is further complicated by the fact that the feared extreme catastrophe can happen at any time, within the several hundred years period.

Experts understand complex, prospective natural and human-made catastrophic incidents differently. Because scientific assessments should be based on the best available science and methods, and because scientific expert opinions justifiably differ, judgments through ranking or voting on how well the architecture will work should be formal and replicable. The overall architecture may, after the fact, be shown to either be weak or sound depending on how the prospective catastrophic incidents materialize. What contributes to its overall strengthening is a general understanding of the whole by those who might be affected. We summarize the means to account for the different forms of uncertainty affecting the triplet *assumptions-models-results*. We define and use uncertainty probabilistically, as it is done in regulatory law. However, we also describe other measures of uncertainty and the methods. It is important to realize that not all uncertainties can be formalized as probability numbers. Forcing uncertain reasoning into a probabilistic straitjacket may be avoided. For example, we introduce fuzzy sets and models such as controllers and Dempster-Shafer possibilities: namely, *beliefs* and *plausibilities*. The benefit of enlarging the envelope of the representations of uncertainty is that causation should be consistent with the different states of uncertainty that characterize its inputs, cause, and effect. The combination of uncertain reasoning with scientific causation answers legislative and policy questions. These are guided by the details provided by regulatory and administrative requirements and are affected by judicial reviews, should these happen. The flow of information in this Book is depicted through the linkages between Chapters via curves labeled *key information paths* to indicate the feedbacks between science and policy:

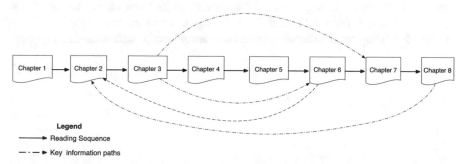

Our science-policy discussions review the main legislative and policy principles that govern the regulation of future catastrophic incidents and the basis for their assessment. One such principle is the EU's Precautionary Principle, PP, as put forth in the EU's Treaty on the Functioning of the European Union (2012), TFEU, interpreted in European case law and by the EU Commission. US federal law has contributed precautionary criteria in health, safety, and environmental laws that have a similar intent to that of the EU's PP. Those criteria and principles have been used, with or without changes, by other countries to justify their environmental, occupational, and safety decisions. We review their salient aspects as well.

The policy concerns are risks, consequences, costs, and benefits in the context of probability (often stated as frequency-magnitude (FM)) distributions and exposure-response damage functions, many of which are multifactorial. The latter of these two functions represents the combined effect of one or more risk factors on the probability of damage caused to those at risk. Damage functions represent physical (in the sense of biological, structural, environmental, and so on) mechanisms that link exposure to damage. These may be known, partially known, or even be conjectural. Data can be obtained from either experiments or observations or both. The levels of the factors to which those at risk are analytically linked, and the damage that they can cause, combine deterministic modeling with statistical estimation. Typically, the historical record is the evidence to calculate frequency and magnitude, at one or more physical locations, consistent with the nature of the catastrophic incident. Often, the historical record contains gaps, outliers, may be censored or truncated. Although we do not discuss in detail their analyses, methods ranging from simulations to data imputation methods can be used to account for some of these issues. For instance, outliers are an important aspect of the analysis and evaluations of catastrophic incidents in that they may be either single *Black Swans* or *dragon-kings*.

Regarding some of our examples, we rely on freely accessible Wolfram *Demonstrations* for two reasons: (i) they are available at no cost and (ii) they can be directly manipulated by those interested, once downloaded, to simulate alternatives to our examples without having to purchase Mathematica® itself. These

Demonstrations fall within the much wider set of Demonstrations that Wolfram Mathematica® makes available and continues to produce. Their benefit is that they extend our examples through other Demonstrations that can clarify many of the analytical methods and concepts that we discuss and that cannot be included in a single Book. The Demonstrations also contain technical explanations and references.

Ravenna, Italy   Paolo F. Ricci

Conflicts of Interest

The author declares no conflicts of interest.

Acknowledgments

The Chapters of this Book have been extensively reviewed by several graduate students in the WACOMA Erasmus Program at the University of Bologna (2019–2021), by two graduate students of the same program (2018–2020) and by Sorin Straja, PhD. Rolf Weitkunat, PhD, has contributed fruitful discussions concerning the aggregation of heterogeneous information, knowledge, and causation. He has reviewed some of my writings in these contexts. Chapters 7 and 8 have been in part supported by PMI, Neuchatel, Switzerland. Dr. Hua Xia Sheng (Xiamen University, PRC) and I have long worked on many of the policy issues discussed throughout this Book. The WACOMA students (2019–2021) have contributed much to the readability and technical content of the final product. I am particularly grateful to them for acting promptly and kindly; many of their suggestions have helped to clarify my writing. Lukas Varnas, a graduate student in the WACOMA (2018–2020) Erasmus Program, and whom I am advising on his MS thesis, has been my research associate. His work, from commenting and criticizing the Chapters to the more mundane setting up this Book in its templates, has been excellent. Some examples and discussion developed in an earlier book, also published by Springer, regarding meta-analysis and Monte Carlo simulations (these sources' copyright rests with this Author) have been modified and included in this Book. I thank the Springer Editor, Ms. Janet Kim, MPH, for her patience and for working with me on this Book. All errors are mine.

Contents

About the Author

Paolo F. Ricci, PhD, LLM is a professor at the EU's Erasmus Mundus Programs in Italy, Spain, and Portugal. Previously he was a professor at the School of Public Health at the University of Massachusetts at Amherst; professor at Holy Names University in Oakland, California, USA; visiting professor at Xiamen University in China; and professor at the University of Bologna in Italy. For more than 30 years, Dr. Ricci – a senior Fulbright scholar (specialist, 2010–2015) and appointed peer reviewer for Fulbright Specialists selection (2013–2014) – has led qualitative and quantitative analyses in public health and epidemiology and conducted experimental work in, among other countries, the United States, Canada, Italy, Australia, France, Vietnam, China, the Ivory Coast, and the European Union. He was the head of the Environmental Technologies Clearinghouse of the IEA/OECD (with full diplomatic status) and has served as a peer reviewer of U.S. Department of Energy (DOE) activities regarding human health risks from past nuclear weapons tests at the Nevada Test Site. Until 2014, Dr. Ricci was an associate editor of the journal *Environment International* for more than 15 years.

List of Symbols

This List of Symbols defines the symbols used in this book. Some of the symbols may have more than one use. For example, $f(x)$ can mean the value of the function, f, for the value x. However, $f(X)$ or $f(x)$ can also symbolize the density or mass function of the random variable X. The text should clarify the context for the symbols.

Symbol	Definition
$\pm$	Plus or minus, alternatively minus or plus
$\mid$	Conditioning; conditioned on
$*$	Multiplication
$\sim$	Is approximately equal to
$\equiv$	Identically equal to
$\neq$	Not equal
$\leq$	Less than or equal to
$\geq$	Greater than or equal to
Δ	Discrete change, an increment or a decrement
$\partial(.)/\partial x$	Partial derivate of the quantity in parentheses with respect to x
∞	Infinity
$\int f(x)dx$	Indefinite integral
$\sqrt{}$	Square root
$\prod$	Multiplication
$\sum$	Summation
$>>$	Much greater than
$\#$	Number, count
$\lvert . \rvert$	Absolute value; number of, count of… …
$e(.)$	Exponential function; $e = 2.71828\ldots$
$\exp(.)$	Exponential function; $\exp = 2.71828\ldots$
s.d., sd	Standard deviation of the sample
$Var(.)$	Variance for sample or population
μ	Population mean

Symbol	Definition
σ^2	Population variance
λ	Rate
E(.)	Expected value
<.>	Expected value
$\log_2(.)$	Logarithm of (.) to the base 2
x^n	x raised to the nth power
$\sqrt[n]{x}$	nth root of x
pr	Probability
F(.)	Cumulative distribution function
f(.)	Density function
H(.)	Cumulative hazard function
h(.)	Hazard function
OR, O.R.	Odds ratio
RR, R.R.	Relative risk
$\int_a^b f(x)\,dx$	Definite integral
$\otimes$	Convolution
pr(Y|X)	Probability of event Y, given X; conditional probability
pr(Y AND X)	The joint probability of event X and event Y
pr(Y OR X)	The probability of either event X or event Y or both
nCr	The number of combinations of n objects , r taken at the time
$d(.)/dx$	The first derivative of the quantity between parentheses, with respect to x
$\ln x$	Logarithm to the base e of the number x
n!	Factorial of the number n
$a \in A$	A is an element of the set A
$f^{-1}(.)$	Inverse of f(.)
$\log x$	Logarithm to the base 10 of the number x
$X \rightarrow Y$	X is an input to Y
$\forall$	For all …
i.i.d.	Independent and identically distributed
dx	Infinitesimal change for the variable X
$\cup, \cap$	Union, intersection
L, M, T	Length, Mass, Time
$X{\sim}N(.);\ U{\sim}T(.),$ $Z{\sim}LN(.)$ …	The random variable is normally distributed, the random variable U has a triangular distribution, the random variable Z is log-normally distributed, …
$\bar{x}, \bar{y}$	The sample means for observations taken on X, Y, and so on
$\propto$	Proportional to
:=	Assignment
$p \rightarrow q$	Implication: proposition p implies proposition q

Chapter 1
Black Swans, Dragon-Kings, and Other Catastrophes: Caught Between Infinitesimals and Googols

1.1 Introduction

A *Black Swans* is a metaphor for catastrophic, extremely rare, and unexpected events (Taleb 2007). The phrase seems due to the Roman poet Juvenal's *rara avis in terris nigroque simillima cygno* ... "rare bird on the land that is very similar to a black swan" (Juvenal reportedly was satirizing women ...). It combines the improbable with its effect. Centuries later, in England and in Italy, Juvenal's phrase *rara avis* became a common expression for the impossible. It was associated with *the Old World presumption that all swans must be white because all historical records of swans reported that they had white feathers.* However, in 1697, a Dutch expedition led by Willem de Vlamingh discovered black swans in Western Australia and the phrase changed to mean that a perceived impossibility might later be possible. In 2007, Professor Taleb was reported by the *New York Times* (May 6, 2007) to say that a *Black Swan* has well-defined aspects. These include being significantly different from the remaining sample of data. Statistically, it is an unexpected outlier associated with a catastrophic outcome. The combination of being an outlier and a catastrophe leads to explanations that may be causally incorrect and are aimed at justifying the lack of ex ante precautionary or preventive actions.

We have informally grouped catastrophic incidents into *Black Swans*, *dragon-kings*, routine, and non-routine. This accounts for different levels of certainty about those events and their consequences. Table 1.1 contains a summary of the types of catastrophic incidents we discuss. The term *incident* implies both cause and effect: the event and its direct and indirect consequences such as cascading events and their consequences, although these cascades are not fully discussed, to keep the discussion simple. Practically, the contents of this book applies to cascading catastrophic incidents from networks of events to their consequences.

An *extreme impact* being *outside regular expectation* is an *infinitesimal* probability of occurring (e.g., probability $<< 1*10^{-4}$ and clearly *outside regular expectation*); the *googol* ($1*10^{100}$) might measure the magnitude of its consequences. An

© Springer Nature Switzerland AG 2020
P. F. Ricci, *Analysis of Catastrophes and Their Public Health Consequences*,
https://doi.org/10.1007/978-3-030-48066-0_1

Table 1.1 Summary of types and characteristics of catastrophic incidents

Type of catastrophic incidents					
Type of event → Key characteristics ↓	Non-routine (NR)	Black Swans (BS)	Gray Swans (GS)	Dragon-Kings (DK)	Notes
Distribution of consequences, given the class of catastrophic incident	Rapid convergence to a large consequence	Slow convergence to an extreme consequence	Slow convergence to an extreme consequence	May include cusps over the domain	Consequences range from aesthetic to ecological; from physical to financial. Measured by a variety of metrics
Before the fact theoretical predictability	High, well-established probabilistic processes	Low, heuristics (e.g., "bubbles") but no warnings	Medium. Theory is available; warnings occur	High. Theory and early warnings are available	Relative to NR
Historical data can improve empirical predictability	High, statistical analyses used routinely	Low, the past does not predict the future	Medium, available historical data	High, historical data are not predictive	Relative to NR
Impacts	From large (> than 10 human deaths) to extreme magnitude				From low to very probability events. From prompt deaths to ecological
Justification of ex-ante precautionary or preventive policies	Based on risk-cost-benefit (RCB) analysis	Cannot be related to RCB analysis	Can be based on RCB analysis	May be related to RCB analysis	Precautionary policies may not be practically implementable due to the magnitude of the instantaneous consequences

Routine events excluded but discussed in the text

example in which numbers combining infinitesimals and googols may be the sudden destruction of New York City by a meteorite. The overall magnitude of the damage to the USA, and to the rest of the world via cascading mechanisms can only be thought of in terms of googols of dollars ($1*10^{100}$ [$]). This event has a computable probability of happening; its effects can be unimaginably large.

More practically, a *Black Swan*, *BS*, appears to be a much rarer and larger magnitude event than a *dragon-king*, *DK*. The latter is the true *outlier* in a distribution of rare events. *DKs* may be predictable from known mechanisms that may be unexpectedly amplified of changed by external risk factors (e.g., forces). The distributions and probabilistic appearance of the consequences from: common, rare, *DK*,

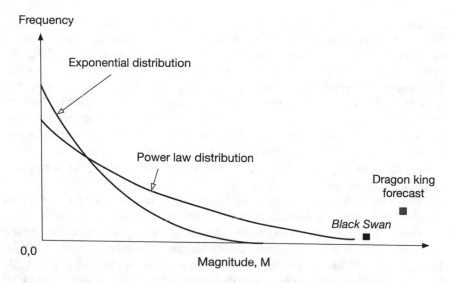

Fig. 1.1 Idealized depictions of dragon-kings, Black Swans, and routine catastrophic events using empirical long tailed power law distributions and an exponential distribution function (Magnitude $\sim$k∗Exp(λM)). The origin of the density and magnitude axes is the point 0,0. *Note*: Data to which each distribution is fit are not shown. Each distribution is empirical, data driven. The DK is a predicted event that would shift the power law upwards, thus thickening the tail, had it been observed

and *BS* events. These are depicted in Fig. 1.1. *DK*s have a different theoretical and empirical (based on data) explanation than *BS*s. Figure 1.1 depicts *DK* and *BS* events. The *Black Swan* is associated with an undisclosed physical causal mechanism. However, it appears to belong to the stretched probability distribution (a *power law* distribution) and thus is theoretically predictable. The power law distribution resembles the exponential distribution function below it; however, it is much longer and has a much thicker tail. The *DK* is depicted as being of larger magnitude than the magnitudes associated with the tail of exponential distribution and outside its domain.

The *DK* is the true *outlier* for power law distribution. In Fig. 1.1, both the continuous distributions are fitted to data that are not shown to avoid clutter. The *DK* may be generated by a transient change that can occur through a positive feedback or from an exogenous factor affecting the mechanisms associated with the power law curve. Hence – unless it is an outlier from transcription or other error – it must be addressed (using both empirical data and theory) so that policy is correctly informed. Following Cavalcante et al. (2013), *DK*s are the result of *organizations of system components that are statistically and mechanistically different from their smaller siblings* such that *the most extreme events do not belong to the scale-free distribution*. Terms such as *scale, centrality,* and other elements of probabilistic distribution functions are discussed throughout this book. If data were available, the *DK* event would be characterized by a sharp spike, preceded by a dip in the otherwise roughly constant negative slope of distribution fit to the data. Moreover, there could be bubbling (a set of up and down data) immediately preceding the actual DK. These two sets of events, characterized by very lower frequency magnitudes, signal of an impending catastrophe, unlike the variability of the data along the power law itself. The *DK* is an outlier – ex ante unknown to exist – that is qualitatively and quantitatively different from the probabilistic behaviors (the data) that precede it and that belong to the power law distribution. For example, Sornette (2002) studied large changes in stock exchange indices and found that *bubbling regimens precede them*. Johansen and Sornette (2000) state that (for runs of losses) about 99% of them follow an exponential distribution with a fat tail.

Regarding the physics of *DKs*, their state-space representation, for example, a coupled oscillator, shows that their trajectories can deviate from what may be considered an average behavior and result in an extreme event (Sornette 2006). This change is not an outlier because it is explained by the deterministic differential equations that generate those trajectories; however, the trajectories of the coupled oscillator mechanism become chaotic. For example, a *DK* may be represented by the *characteristic earthquake hypothesis*, which occurs when coupling is high. Large magnitude earthquakes may be distributed as power laws for specific physical regimens of relaxation and stress build-up. Amplification, perhaps by an additional unexpected mechanism, leads to catastrophe. The large forces may not cause catastrophic incidents, but their amplification does. Exogenous stresses are also important because, once an unstable equilibrium is reached, the system is at a critical point: a very small force breaks the proverbial camel's back. A signature for that event, such as bubbling, initiates with *hot spots* within the chaotic attractor. It follows that the nature and length of the *window* of observations is an important empirical aspect of any assessment*s* of catastrophic incidents, provided the data exists.

Regarding *Black Swans*, in an interview with Professor Taleb the interviewer (Wikipedia contributor 2014) stated, "Far more problematic are things that appear perfectly Gaussian in small samples but have rare extreme jumps. For example, ... worldwide influenza mortality rates over the last 300 years leaving out 1732, 1781, 1802, 1830, 1847, 1857, 1918, 1957 and 1968, you could conclude you were looking at a Gaussian distribution. You could study 30 years of consecutive data without a clue that global influenza pandemics are possible and therefore understating the mean and wildly underestimate the probability of extreme events. ... Most of the examples in your book appear to be things of the first type, which only require improvements in statistical practice to address. But most of the discussion concerns things of the second type (namely, Black Swans) for which statistics can offer no help. Are you mixing apples and oranges?"

Empirically, the effect of a rare but omitted magnitude from a sequence of historical magnitudes can be devastating to statistical predictions. This raises the theoretical question: Is the unknown *Black Swan* event of concern truly unknown, or is it knowable? In our work, an *analysis* is an explicit, formal, and verifiable system of premises, rules (e.g., the causal model), and conclusions (i.e., results). The analysis of extremely rare events can seldom be deductive that is, premise and rule are known, but not the results. Their analyses will generally require inductive reasoning: the rule (i.e., the *If...Then* formalizing of physical mechanisms) is not known, but the premise (or assumption) and the outcome are known. The second issue is that data for rare events, if they exist at all, inevitably form a *small sample*. If the reason for analyzing extreme events is predicting their time to arrival, it may be incorrect to *extrapolate* from a small sample of results. The reason is that the nature of rare events – although their consequences also occur when routine events occur – has different causal mechanisms. A statistical argument based on probabilities alone may not provide sufficient evidence to differentiate between them.

Predictions of *Black Swan* events seem to create a *chicken and egg* problem.[1] One tentative answer is that a method to analyze such events suggests assessing if multiple lower intensity shocks can predict failure (e.g., cumulative damage leads to failure). This suggests that the mechanism (e.g., the cumulative effect of a sequence of forces of smaller magnitude) is known. We partition prospective *Black Swans* into: (i) *immediate Black Swans* A; and (ii) *delayed* Black Swan B, Table 1.2. This qualitative partitioning can be further refined by increasing its granularity from a 2*2 to j*k. The partition in Table 1.2, bolder box, has low granularity based on time to occurrence.

In this Table, *Black Swans* are further classified according to the imminence of the threat posed USGS (2015):

1. *Immediate Black Swan* – An example would be an asteroid predicted to hit the Earth within few days. The probability is very near one: at best, preventive action may be taken to change the trajectory of the object or otherwise avoid the collision. Consensus is the better form of analysis. The aftermath to the devastation in the zone of impact and the rest of the world is measured in googols of dollars; after the fact, dollar-based measurements may lose meaning. Even if a probability is computable, it is practically irrelevant USGS (2015).
2. *Delayed Black Swans* – Actions can be taken ex ante the catastrophic incident. An instance is the eruption of a mega-volcano. Although there is nothing that can be done to prevent it, a country can prepare for its occurrence because the time to arrival is knowable. Imagination and foresight combine with modeling to bind to the magnitude of the probable consequences. Precautionary actions ex ante the catastrophic incident can be developed. Policies, procedures, checks, enforcement, and punishments in the case of violations may prevent some damage to sectors vital to society. The fundamental requirement is correct analysis, legislation, regulation, and control.

We discuss routine and non-routine catastrophic incidents, *dragon-kings*, and *Delayed Black Swans* through causation between the hazardous event and the probabilities of the consequences from that single event. Although scientific opinions of causation may conflict, any assessment of choices would have to be based on cause and effect USGS (2018). The causal context attempts to forecast or predict plausible

---

[1] If the egg came first: no organism can change its species during its life, only a mutation can. If the chicken came first, the egg is only maternal. (D.Waller, The Chicken and the Egg, *Mind*, 107:851–8544 (1998)).

Table 1.2 Rough partitioning of Black Swans into immediate and delayed events

Actions before the event occurs	A: Immediate Black Swan	B: Delayed Black Swan	Comment
Prediction, control, enforcement, between the many alternative actions.	*No*	*Yes*	For B modeling can be done: causation may be ascertained over time through quasi-Bayesian methods.
Consensus on anticipatory actions.	*Yes*	*No; voting and ranking is critical*	For A there may be some modeling via heuristics.
Uncertainty	Probability of event can be very high but the event may have lowest predictability.	Probability of event can be very low.	Magnitude of the consequences is *knowable* for both A and B.

A similar partitioning can be made for other types of catastrophic incidents. The outcome space implies both immediate and delayed consequences

policy answers to the question: How can the catastrophic event occur and what is the prospective magnitude of its harm? The answer should lead to the preferable action by decision-makers regarding what can be done about it before it happens. Historical events over time (or over an area) make their data informative, but this is after the fact information and knowledge, *K&I*. For example, *stress tests* and other simulations can approximate the actual behavior and failure modes of a system. A major difficulty is to account for its often-multifactorial causal factors, mechanisms, and their uncertainty. A mechanistic explanation describes the behavior of a system prior to its catastrophic failure in terms of its causal variables and paths connecting them and leading to the feared result. The conditions that we seek to analyze are summarized by the probability that *Black Swans* exist, given that we know with certainty that white swans exist and that some white swans are *white swans*. The analysis of causal choices should serve as a warning regardless of the action eventually taken by a decision-maker. If society were correctly informed, recriminations may be reduced and be less opportunistic. A way to minimize recriminations and enhance the accuracy of the *K&I* that informs the stakeholders in the *assessment-choice-decision* process. It should also account for the following:

1. Willingness to ignore parts of the existing evidence
2. Incorrect representations, analysis, and conclusions
3. Cognitive errors and biases

Example *Waterborne arsenic poisoning in Bangladesh* – Arsenic is endemic in the flood plains of the Ganges River, affecting India's West Bengal and large parts of Bangladesh. Arsenic occurs from either oxidation (of arseno-pyrite) or oxyhydroxide reduction (via desorption from iron compounds). The *dragon-king* catastrophic incident was water contamination from farm activities (e.g., feces from animals reaching the surface water untreated) that caused deadly diarrheal diseases. To prevent these diseases to occur, a UNICEF effort in the 1980s provided for a large number of shallow water wells to be installed in villages. UNICEF achieved >80% *safe* drinking water by 2000. Although arsenic contamination of the ground water had been reported (in 1993), the UNICEF did not account for this contamination of the shallow aquifer from which the wells it installed drew their water, although it was known to be endemic to the Ganges' delta. Arsenic is a known carcinogen (at low doses it causes bladder, skin, and lung cancers), but – for Bangladesh and West Bengal – the catastrophic consequences occurred at higher (toxic, rather than carcinogenic) exposures. After ingestion, arsenic causes severe skin lesions that lead to gangrene, skin cancers, neurological effects, cardiovascular diseases, hypertension, and diabetes mellitus. The number of people at risk of such environmental exposure is astounding. The World Bank (2007) estimated that approximately 20 million citizens of Bangladesh are affected by arsenic poisoning, with another 1.5 million in West Bengal (India). Ignoring early warnings and best practice, the British Geological Survey, BGS, in 1992 did not test for arsenic contamination of the ground water from which the wells drew water for public use. This resulted (in 2003) in a civil law case against the BGS (in the UK, by a Bangladesh citizen) that went all the way to the House of Lords (in 2006). There, the plaintiff's lawsuit was rejected. A newer solution was proposed: deep wells. But these are costly; the areas at risk of poisoning did not have a water supply infrastructure that could provide potable water from the deep wells. This example underscores the fact that prospective choices involve risks-consequences tradeoffs due to limited funds and partial knowledge. Here, a known diarrheal disease from a human-made source overwhelms probable gangrene and other diseases caused by exposure to arsenic. In other words, the evident trumps the suspected.

Modeling reconstructs cause, given an effect and suggests the question: Is there a potential irreversible barrier such that the consequences do not allow reconstructing the event that generated them? The question is exemplified by (Wikipedia 2015): "…a melting ice cube …. If you see an ice cube, it's easy to predict the shape of the puddle that will result when it melts. But if you see a puddle, it is impossible to determine the shape of the ice that created it."

The ice cube analogy involves known physical mechanisms: their dynamics (changes over time) and thermodynamics (heat exchange). Yet, observing outcomes *may not* help build the theoretically correct model. Unless there are theoretical reasons for choosing a specific type of model, and there is a small number of parameters relative to the amount of data, modeling can be as difficult as trying to guess the shape of an ice cube from observing a puddle of water. We know both but the processes cannot be deduced from the observations (e.g., a ball of ice turned into the pool of observed water. The heat equation governs ice cube melting: it is a well-known problem that can be solved forward but is ill-posed for backward solution. The consequences of catastrophic incidents may be understood by examining the present – a forensic analysis. From system identification methods to Bayesian model averaging, researchers and practitioners have been concerned with using observations to build competent statistical models of the underlying reality and reduce its complexity via data reduction methods. However, aside from scientific issues, reconstructions can induce ex post facto justifications. Complex and not well understood mechanisms, some of which can occur at the same time, may affect predictions: the *perfect storm* may happen without apparent warning

Example A *dragon-king* event is the NASA's *Challenger* shuttle disaster (Presidential Commission 1986). The prospective probability of that failure occurring was estimated by NASA to be about 1/100,000 launches. Yet, it occurred on the *25th* launch and killed the entire crew of seven. The miscalculation was due to assuming independent intervening events that lead to the catastrophic failure, the *top event*. Their multiplication yields the approximately 1/100,000 probability of catastrophic failure. This number is much smaller than it should have been from historical data: between *1/25* and *1/50*. The initiating event was an unseasonal nighttime freeze that made an *O*-ring brittle and thus prone to break under the stresses associated with launch, as it did shortly after the lunch took place.

The next example suggests an important class of high probability and large magnitude routine, recurring catastrophic incident. They are routine single events that, when aggregated over a large area in a year, match in magnitude but with certainty the probable catastrophic consequences from natural catastrophic incidents.

Example Mistakes made in hospitals result in the tens thousands of yearly premature and unexpected deaths from a variety of errors. The IOM (1999) high estimate of those deaths was that *98,000* people die per year from errors in medical management; the low estimate was *44,000* per year. Medscape (2019) adds, "Unfortunately, the IOM numbers, shocking as they are, probably underestimate the extent of preventable medical injury, for 2 important reasons. First, they are based on data extracted from medical records. Many injuries, and most errors, are not recorded in the medical record, either by intent or by inattention, or, more likely, because they are not recognized."

1.1.1 Discussion

Empirically, all swans in Europe were once known only to be white. In other words, given white swans (*Cignus*), the existence of black swans is the theoretical alternative hypothesis to the hypothesis that *all swans are white*. Before having observed the color of European swans, by hypothesis, the random variable *swan color* has two probable outcomes: white (*yes* or 1) and black (*no* or 0). These unambiguously define simple outcomes. What is being assessed? If the existence of swans of different colors is being assessed (although white is not technically *a* color), the problem as follows. In a state of the knowledge of only white swans, the existence of black swans is given probability zero – the strictest understanding of Juvenal' *rara avis*. The black swan alternative is assigned zero probability mass. The entire probability mass (100%) is assigned to white swans because these have been observed for millennia, at least in Europe. Next, we consider the probability distribution of the whiteness of white swans. The less white ones fall in the left tail and the much whiter ones fall in the right tails of the distribution of white swans. We can arbitrarily place a *whiteness* bound on the distribution of white swans, namely, a threshold that separates the right tail (e.g., much whiter swans) from the main distribution of what may be labelled *regular* white, and similarly for the equally much fewer and darker white ones. The analysis would require dealing with wave lengths and other complications. Let us change the attribute from color to mass (i.e., weight, in kilograms). *Black Swans* and black swans would still not belong to the sample space of white (and *White*) swans. For example, if we were concerned with very heavy and very light white swans, the distribution of the mass of white swans on either side of the median – and several standard deviations away from it – consists of very few, very light weight, and very few heavy weight *white swans*. The swans with low probability mass are not *Black Swans*: they may be labeled *white swans* to emphasize their being rare. Their very low or very large masses are the rare events. If we knew the skeletal characteristic of *swans,* we could calculate the minimum and maximum theoretical weights of those birds, remembering that mass cannot be negative.

If we decide to use the label Black Swan to represent an exceptional event within all other occurring homogeneous events, it would be theoretically predictable, but not empirically demonstrable, before its actual observation. This Black Swan's key physical characteristics are as follows: (i) being rare; and (ii) having different or remarkably altered mechanisms that produce it. It distinguishes itself from the rest of the events in the distribution by having exceptional evidentiary value. We seek just enough accuracy to be able to predict, with the degree of confidence consistent with the use of the K&I generated, this event. It is the point consisting of low probability and large magnitude that is of concern under precautionary policies. Outliers may not belong to the empirical distribution being used. They may indicate a different mechanism than assumed: they are *sentinels* of another phenomenon perhaps yet to be understood. Alternatively, a white swan outlier is a White Swan that has a specific characteristic – such as weight – that makes it rare within the probability space of the weight of all white swans. Situations such as those enumerated below may produce outliers:

1. An error of commission (transcribing the wrong number) may have been committed.
2. The outlier is real and belongs to the population: it is several standard deviations away from the mean. If so, the researcher can investigate its nature, explain it on appropriate empirical and theoretical grounds, and then include it in the analysis.
3. The outlier is real but belongs to a different population. In this case, the researcher should omit that observation but report that omission and its implications.
4. There are several outlying observations. It is probable that two different phenomena are measured in the same sample. Physical and other explanations may be used to describe and account for them.

Outliers, without further statistical tests and changes to the methods for estimation, increase the variability of the central value of an estimated parameter: estimation is inaccurate if the outlier is erroneous.

The magnitude of catastrophic events can be either under- or overestimated, depending on the empirical distribution of the data used. If the theoretical distribution is wrong, then the model to describe those events and their probabilities is also wrong. If we can only rely on a sample, the empirical tail or tails depends on the data and the protocol used to obtain them. The context of the assessment should dictate the window of observations (for time-dependent events), the type of sample to be taken (e.g., random or purposive) and accounting for both Type 1 and Type 2 errors (i.e., false positive and false negatives), as well as the choice of an incorrect model leading to inaccurate causation. The probability law that generates the data may have a theoretical and physical basis but cannot account for missing risk factors, if those are relevant to causation. The record of events and their probabilities rank-ordered by small to large magnitudes may – but does not assure – a causal relationship such as exposure-response. For example, the linear no threshold at low doses, LNT, dose-response model is a statistical conjecture that may not be resolved by increasing the sample size due to cost and feasibility of the required experiment.

Understand the complications in modeling of catastrophic incidents of any type for prediction and for public policy requires some level of completeness of representation. We provide some aspects of completeness for a catastrophic incident as follows:

1. *Definitional completeness*: the set of formal definitions.
2. *Modeling completeness*: the best available science, mathematical, and statistical models.
3. *Empirical completeness* and *boundary completeness*: consistency with scenarios that may be unusual a priori but that, ex post, are recognized to have been predictable from historical or other studies that were neither used nor considered.
4. *Information and knowledge completeness*: available studies that yield both positive and negative results.
5. *Defaults, bias,* and *representation completeness*: assessment of defaults, probabilistic representations, analysis of biases, use of the appropriate measure to represent non-probabilistic uncertainty (e.g., randomness, epistemological uncertainty, fuzzy sets).

Completeness also requires having the budgetary means to achieve it. The value of additional information (e.g., as the value of sample information, VOI) and that of alternatives choices should be assessed according to the appropriate criterion (e.g., select the choice that *min or max expected value*; alternatively selected the choice that is the *max-min*. The process defining a prospective reality, its modeling, and the assessment of choices consists of several aspects, some of which are included in Table 1.3.

Table 1.3 Empirical and modeling aspects of the completeness of assessment

Aspect of completeness of analysis	*Black* swans	Routine, non-routine and DKs	Comments
Outliers	Empirically detectable	Yes	May be *sentinel* outcomes
Model choice	Conjectural	Theoretically and empirically possible	Present as an issue
Model specification	Limited	Actual risk factors may be known	Present as an issue at low levels of exposure
Bias-variance tradeoffs	No to relevant	Particularly relevant	Combines with model specification
Cognitive bias	Present	Possible	Present
Biases (e.g., recall)	Present	Possible	Present
Rational choices based on theoretically sound criteria	Possible	Exist	Critical for justifying choices, but may be dictatorial
Answer the wrong question	Yes	Yes	Possible for both, but importance decreases as time to catastrophe increases
Ask the wrong question	Yes	Yes	Possible for both, but importance increases as time to catastrophe decreases

1.1.1.1 Accident Analysis: Simple Illustrative Calculations

We exemplify some analysis of routine accidents events based on travel data. The calculations give an insight on how physical units are used; all number are hypothetical and kept small to facilitate back of the envelope calculations.

1. In a country there are *5000* prompt accidental deaths, per year, due to car crashes. Let the yearly total number of kilometers driven be $1.5*10^6$ in that country. Then: 5000 [deaths/year]/$1.5*10^6$ [km/year] = $3.33*10^{-3}$ [deaths/kilometer traveled].
2. Suppose that the average number of kilometers driven per day is *10* and that the number of driving days per year is *300*. Then: $3.33*10^{-3}$ [deaths/kilometer traveled]*10 [km/day]*300 [days/year] = 9.99 [deaths/year].
3. Assume that a monetary loss associated with average death is $100,000. This is not the value of life, rather it is set of costs associated with contingent expenses by the survivors. Then: 9.99 [deaths/year]*100,000 [$/death] = 999,000 [$/year].
4. Suppose that there are 5000 deaths per year due to car crashes and that the resident population is 20 million people. Then: 5000 [deaths/year]/20 [million residents] = $2.5*10^{-4}$ [deaths/resident-year]. If this number is multiplied by the expected remaining life, say 60 years from the 16th birthday (the assumed legal age for driving), then: $2.5*10^{-4}$ [deaths/resident-year]*60 [years of expected lifetime at birth] = 0.015 [death/resident].
5. Calculate the reduction in life expectancy due to prompt death from a car crash in the country. Assuming that an age- and sex-specific life expectancy is 30 years, then: 0.020 [death/resident]*30 [years] = 0.60 [death/resident-year].

6. Assume that there are 10,000 yearly deaths due to car accidents in country XYZ. Assume a general population N of 50,000,000 persons in that year. Assume that the average driver drives 10,000 km per year and that there is 1.00 [person/car-year]. Then: What is the number of deaths per kilometer, in the general population of country XYZ, for that year? The answer is: 10,000 [cause-specific deaths/year]$*$1/50,000,000 [1/N]$*$1/1 [person/car-year]$*$1/10,000 [car-year/kilometer driven] = $2*10^{-8}$ [cause-specific deaths/kilometer-year driven].

7. Suppose that a city has a population N of 100,000 persons in a year and that a proposed activity can raise the individual background yearly cancer risk by $6.5*10^{-6}$ in that city. Then: Expected annual cases$_{city}$ = $(6.5*10^{-6})*100,000$ [N/year/activity] = 0.65 [N/year/activity].

8. Suppose that a district in that city has a population of 1000 [N], that the individual increased yearly risk is $1.0*10^{-5}$ and that those persons are exposed for 15 years to the hazard. Then: Expected cases$_{district}$ = $1.0*10^{-5}*1.000$ [N/year/activity]$*$15 [years] = 0.15 [N/activity].

1.2   Uncertainty and Variability

Uncertainty and variability describe two different states of incomplete knowledge from the state of being without doubt. *Uncertainty* in a natural language such as English describes something that is akin to being *probable*: a state of knowledge expressed as a non-quantitative judgment that something can occur. Alternatively, uncertainty may describe something that is *possible*: an even lower state of certainty. Being uncertain suggests that several reasons may cause that state to be present. *Variability*, for example, the variability of a specific characteristic, given data in a sample or a population, may be due to the heterogeneity of the individuals with the specific characteristic. This variability is different from the variability of a model's outputs, if the model does not account for the factors that characterize the mechanisms generating the output. When so, its predictions are theoretically and empirically questionable. As the NAS (2004) states, in the context of the Clean Air Act, "uncertainty analysis is the only way to combat the 'false sense of certainty,' which is caused by a refusal to acknowledge and (attempt to) quantify the uncertainty in risk predictions." For the NAS, "uncertainty is the lack of precise knowledge as to what the truth is, whether qualitative or quantitative; it is characterized by lack or incompleteness of information." Uncertainty, as the NAS suggests, *depends on the quality, quantity, and relevance of data and on the reliability and relevance of models and assumptions.* Variability is defined as *the true difference in attributes due to heterogeneity or diversity.* This Agency also adds that *variability is usually not reducible by further measurement or study, although it can be better characterized.* The methodology that ensues from these considerations has been called *quantitative uncertainty analysis.* As this agency suggests, it is unquestionable that regulatory modeling *assumptions and default are unavoidable as there never is a complete data set to develop a model.* A typical default is the linear at low doses, no threshold dose response model, LNT, used to assess the magnitude of the consequences from exposure to chemical and ionizing radiations conjectured to result in a variety of cancers. This default causal model is generally – but not always – incorrect (Ricci and Tharmalingam 2019; Clewell et al. 2019). In this book, we use *probable* to describe an event or statement to which we can attach a probability number that expresses our subjective or objective (in which case probability is a relative frequency) assessment of the degree of certainty about the truth. Certainty implies frequency or probability that equals 1.00; no certainty equals 0.00. When dealing with experimental, observational, and other empirical results the term most often applicable is *frequency*. Simply put, researchers and others think that, primarily, there are two types of probabilistic uncertainty. These are (i) *aleatory* (random variations in the outcomes related to a physical event, measured by probabilities or frequencies and often exemplified by a mechanism such as the toss of a fair coin or a thumb tack); or (ii) *epistemic* (contingent on empirical knowledge with probabilities or frequencies serving as weights for the evidence of that knowledge). *Aleatory* uncertainty can only be better estimated by improving the assessments' methods. We are concerned with epistemic uncertainty. For example, a linear model of dose-response for cancer can be used to assert that the risk of cancer, at a specific dose, is a probability (e.g., $(pr) = 0.00030$). More research (e.g., observational, experimental, and theoretical) can reduce *epistemic* uncertainty making the results more certain. For example, the probability of a binary outcome, such as the toss of the proverbial *two*-dimensional coin, will tend to stabilize around *0.50* as the number of tosses (or *trials*) becomes larger and larger. Of course, it can happen that *100* tosses of a fair coin yield *100%* frequency of heads (e.g., 100 heads in *100* tosses of the the same *rounded* edge coin and a frictionless tossing mechanism that operates without failure and that does not need maintenance). More generally, regarding public policy, the NAS (1996) states that the assessors (of risk) bears: "… the responsibility to use whatever information is at hand or can be generated to produce a number, a range, a probability distribution – whatever expresses best the present state of knowledge about the effects of some hazard in some specific setting."

Probabilistic methods have become part of standard regulatory work. For example, the US EPA has developed a policy for probabilistic analysis done in the context of risk assessment (EPA 2000). To understand its basis, consider the distribution of concentrations of arsenic in the ground water at a specific location. Assume that the data is distributed log-normally. The cumulative distribution function, CDF, of doses of arsenic (in mg/kg of body weight of the average individual at risk) can be used to limit exposure by *policy*. The regulator sets a specific probability (e.g., $1*10^{-6}$, a small individual lifetime probability of cancer attributable to exposure) that, through the dose-response model, yields the corresponding legally acceptable dose. The appropriate public policy, for instance an occupational health regulatory standard, determines the choice of that probability number. In this instance, regulatory policy has strong foundations in public health risk assessment but can be affected by untested aspects of completeness, such as those listed in Table 1.3. One element of completeness is the probabilistic and theoretical independence between the risk factors that cause a specific adverse outcome. It can affect the *tails* of the resulting output distribution, given the distributions of the inputs and the operations performed on them. That is, the tails can be either too fat or too thin, relative to what they should be if no correlation were correct. The issue is the unstated desire, adopted in practice, to avoid having to assess correlations between risk factors. The form, calculations, and probabilistic properties of complicated correlations would make the analysis much more difficult. Some of the practical implications of this aspect of completeness are as follows. The estimates of central tendency, such as the median or mean, of the output distribution are either unaffected or mildly affected by correlations. However, importantly, the tails of the resulting distribution *are* affected. This result must be assessed because assuming independence can mask the true joint variability: a key concern of precautionary policy. In the simplest instance, if the assumption that two or more independent input random variables are uncorrelated is not true, and the dependence measured by the coefficient of correlation, is positive, then the *tails* of the output distribution will *underestimate* the true probability. If it is negative, although the true risk can be negligible, it may be overestimated.

The description of a complex catastrophic incident (e.g., using a causal network with feedbacks and multiple variables) requires representing the uncertainty that affects each of these components and how propagating and fusing them affects their aggregate output. For example, in this book we will be concerned with the following situations where:

1. Frequencies may not be appropriate for some rare events because those frequencies are not empirically known; degrees of belief (e.g., subjective probabilities) may be necessary.
2. Probabilities may be crisp numbers, given the information on which probabilities conditioned. Distribution functions describe the relationship between crisp numbers and their probabilities.
3. Probability intervals (such as imprecise probabilities which contrast with crisp probabilities) may be more appropriate, but these also require some level of K&I.
4. The amount of K&I may be large for outcomes around the center of a distribution but may be much rarer in the tails or tail of that distribution. In this case, upper and lower probabilities (Dempster-Shafer beliefs and plausibilities) may be more appropriate.

Given these situations, deterministic choice may be difficult to justify. Choosing a deterministic value without statistical guidance may imply knowledge that is questionable and possibly biased. If the numbers available form a sample, we are on surer grounds because statistical theory guides the analyses. For example, if we want a number that is the most likely to occur, in some cases the best choice is the mean of the population, μ_x. It is demonstrably *best* by the maximum likelihood principle. In general, more information leads to greater value relative to decisions made with less information.

Many choices are based on qualitative consensus rating the importance of attributes and their levels. For example, suppose that a rating system is used to compare two different options, A and B, to determine which should be ranked higher. If the overall rating of the risk is to be based on component ratings of several risk components (or factors), then: How should the overall risk rating of alternatives A and B depend on the component ratings? Some properties include the following:

1. Which of alternatives A and B is rated higher in the overall risk rating should depend only on their components' ratings. The components used to rate risk should be sufficient to do the job: together, they determine whether A is assigned a higher, equal, or lower rating than B.
2. Which of A and B is rated higher on overall risk depends on each of their components' ratings. Specifically, if A and B are identical in all respects except that A rates higher or worse than B on one factor (e.g., exposure), then B should not be rated higher than A in the overall risk rating. This property should hold for all the risk components: none of them is irrelevant.
3. If A rates higher (or lower) than B on every component rating, then B should be rated no higher (or lower) than A in the overall risk rating.
4. Risk ratings of A and B should be based only on their own data, that is, whether A is rated higher or lower than B should not depend on what other alternatives (other than A and B) are also being rated, if any.
5. If one or more component ratings are zero (e.g., for exposure potential or for human health impact potential of exposure), then the overall risk rating should be zero (or "Negligible" in systems with that category).
6. If the rating for a component is uncertain (e.g., if it has a 0.2 probability of being "L", 0.5 probability of being "M", and 0.3 probability of being "H"), then the single "equivalent" rating assigned to that component (i.e., H, M, or L after considering its uncertainty) should not depend on the ratings assigned to the other components.

Rule-based decision procedures – most legal commands we deal with are of this type – that do not explicitly identify or optimize the quantitative impacts of recommended interventions and classifications run the risk of triggering pre-specified actions (e.g., interventions to withdraw or restrict exposure) that unintentionally do more harm than good by creating unintended adverse consequences. Rational risk management requires comparing the probable consequences of alternative risk management actions and then choosing the available action with the most desirable probability distribution of consequences. Substituting *importance* in public health, or other non-consequential criteria, for actual human health consequences as a guide to risk management decision-making, may lead to recommended actions that create more harm to human health than they prevent. Methodologically, no small number of qualitative labels for risk and its components can suffice to make effective risk management decisions.

1.3 Probability Distributions Represent Uncertainty

The key component of uncertainty analysis in the context of regulatory science-policy is the probability distribution functions, pdf: f(t). The argument t means time and will generally be represented by x or X to generalize the discussion. In most analyses, what is used is the cumulative distribution, CDF, F(t) or the *complement* of the CDF: 1-CDF = 1–F(t) of the random variable *time*. The frequency-magnitude curve, FM, is 1-CDF(M), where magnitude can be any appropriate quantity such as number of deaths, loss in dollars, and so on. Other commonly used functions are the hazard function, h(t) and the survival function in which time is the argument of the PDF, rather than magnitude. Table 1.4 summarizes, as a hypothetical and discrete set of probability masses (rather than density) of the consequences: pairs of f $\in F$, m $\in M$.

Example Using O'Hara Wolfram Demonstration http://demonstrations.wolfram.com/ReliabilityDistributions/ the distribution of failures modeled by the log-normal (LN) probability distribution function, LN(f(t; μ, σ)), namely LN(t; 0.79, 0.93). The hazard rate (i.e., the instantaneous failure rate) is h(t) = 1–F(t); Fig. 1.2. The CDF of the lognormal distribution is not shown.

Figure 1.3 depicts a power law-like result obtained by using the log-normal distribution, which is an exponential that, when plotted on a semi-log graph, is linear. As shown, it is an *empirical* distribution. Maclachian's Wolfram demonstration (http://demonstrations.wolfram.com/PowerLawTailsInLogNormalData/) can be used to illustrate how the upper tail of log-normally distributed data can appear to conform to a power law when plotted on log-scale. In this example, one thousand randomly generated points were drawn from a log-normal distribution and used in a log-linear regression, log size = f(log rank): the purple line with negative slope, which we depict in Fig. 1.3. The mechanistic basis implies that the process generating the data has been operating for a sufficiently long time, and that what is observed is the outcome of a proportion of values that are independent from each other. For example, if successive dilutions of chemical are independent, over a relatively long period of time the observed concentrations are log-normally distributed. These laws also apply to economic variables such as income and wealth.

The writings of Taleb (2007) about *Black Swans* suggest that FM representations may be incorrect in part because these functions are generally shown to intersect the magnitude axis. In other words, an FM curve depicted as A in Fig. 1.4, can be incomplete (does not intersect the magnitude axis) on the right side of the diagram, in sharp contrast to what is shown for B, C, and D. These three curves intersect the

Table 1.4 Hypothetical combinations of probability and magnitudes – probabilities do not add to unity: the set of (m, f) depicted is limited to events from 10^{-2} to 10^{-6}

	Combinations of frequencies and magnitudes			
Magnitude, m	10	250	*1200*	*3500*
Probability mass, f	10^{-2}	10^{-3}	*10^{-4}*	*10^{-6}*

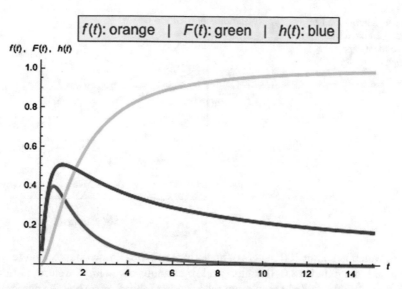

Fig. 1.2 Density function, f(t), cumulative density function, F(t), and hazard function, h(t), for the log-normal distribution LN(0.79, 0.93), developed with O'Hara http://demonstrations.wolfram.com/ReliabilityDistributions/

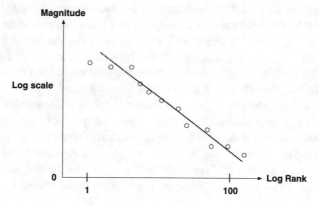

Fig. 1.3 Log-linear regression, log size = f(log rank) of a power law set of data and fitted regression line

magnitude, M, axis indicating much more complete knowledge. This depiction is further discussed in this book; for now, it may represent the opinions of four experts, one of whom can only commit to providing her partial knowledge, unlike her three other colleagues.

The issue of the unknown upper values of the magnitude M, in the F-M plane, is critical when assessing catastrophic incidents and their consequences. Is it true that the upper bound is theoretically or empirically not knowable? We develop two situations. In the first, suppose that a family of 5 decides to buy a 1.5 million dollars

Fig. 1.4 Alternative FM curves generated by independent experts through elicitation

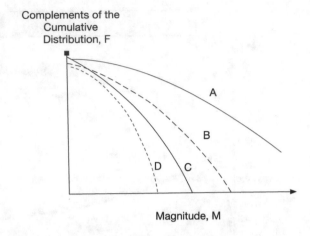

Complements of the
Cumulative
Distribution, F

Magnitude, M

home, which is within 100 meters of an active (and thus known) fault. That fault has generated magnitude 8.0 (Richter scale) earthquakes with a return period of 200 years. What could the maximum magnitude of the damage be? We can hypothesize two upper bounds. The maximum consequences is the destruction of the family and their dwelling as the result of the earthquake. Alternatively, if the family were having a wedding party with 100 people in the house, then the maximum toll would be 100 casualties. Consider an earthquake with even lower probability but much larger 9.5 (Richter scale) magnitude. Suppose that it can happen in an area with a nuclear power plants, near an ocean with a bathygraphy that could result is tsunamis about 20 meters high, should that earthquake happen. Let the number of people at risk be 10,000,000. The magnitude of the probable damages can be approximated. If so, there may be a plausible upper bound on the total number of deaths, property destruction, and collateral damage under either event. It is a matter of simulations and analysis of ex ante choices to develop alternative solutions and decide on the preferable one. As Fig. 1.4 depicts, we *can* approximate upper bounds on most catastrophic events.[2] Some experts might prefer to withhold judgment on the maxima for non-technical reasons. As Fig. 1.4 depicts, *A*'s scalable distribution might correctly characterize the probability of extreme events, relative to the alternatives.

1.3.1   Comment

Mandelbrodt (1998) distinguishes between *benign* from *wild* data, strongly suggesting that the former do not describe reality, at least in most instances relevant to catastrophic events. A stable reality, as seen from the benign process, is the result of the law of large numbers and the central limit theorem. The former results in a probabi-

---

[2] Should the plot be semi-log (log of frequency *v.* natural value of magnitude) the curves would look very different: the Gaussian would actually appear as straight line, while the scalable distribution would be nonlinear and as a flat *U*-shape, convex to the origin.

listic limit (the expectation); this limit is normally distributed; moreover, past and future observations are asymptotically independent. The latter of the does not, at least in the traditional sense of the central limit theorem, CLT. In some cases, these limits are approached very slowly. The *wild* case characterizes processes the outcomes of which, over time, have very long and slow cycles where the amplitude of a frequency (*f*) for the errors is inversely proportional to the frequency itself. In this case, neither the law of large numbers nor the central limit theorem applies. A reason is that – regardless of the time scales – changes in the physical structures that generate the observations in the past can reoccur in the future, except that the recurrence is not foreseeable. Another reason is that fluctuations in the data should not be assumed to be *benign*: they have to be demonstrated to be so. For example, the Cauchy density function is a distribution with *uncommon* characteristics. Each Cauchy random variables, $X(x)$ with $x \in X(x)$, has a density function $[\pi(1 + x^2)]^{-1}$, its expected value also is a random variable with infinite value. It follows that the expectation of this distribution cannot be understood the same way as the mean of a normal or other common distribution. In another instance, the log-normal distribution of the random variable $X(t)$ has a hybrid characteristic: as time tends to infinity, it has finite moments; but if the interval of time is short, then these quantities become unstable and the distribution is *wild*. Changes in the period (or *window*) of time over which the data are observed can entail an unexpected behavior. The combination of length of the natural cycles, data gaps, incomplete knowledge, and budget constraints affects how we should assess catastrophic events. The net effect suggests that analysts may look at a rare event as belonging to benign distributions.

As discussed, power laws are summary descriptions of reality. Typically, a power law (e.g., $f(x) = kx^c$, or $g(x) = Kx^c[h(log(x)/log(a)]$, in which $g(x) = g(1 + x)$ accounts for periodicity, is scale invariant. This means that the scale at which the function is observed does not affect its shape. For example, it does not require a reference object (such as a ruler next to the object to indicate its scale) to gauge its scale when it is shown. When plotted on log-log axes, power laws are linear. It is essential to understand mechanisms, statistical, and empirical aspects because power laws are associated with infinite systems, while real systems are finite (Stumpf and Porter 2012). A classic example is allometric formulae in which empirical and physiological aspects occur over several order of magnitude of the mass of mammals. Thus, allometric formulae are a synthesis of reality that lacks microscopic causation: they are empirical, macroscopic regularities that can be modeled by power functions. However, Willinger, Alderson and Coyle (2009) and Stumpf and Porter 2012 find that the evidence for power law is often negligible because the statistical assessments of these regularities are incorrect and the power law can be spurious (Stumpf and Porter (2012). Power laws are distributions characterized by a single very long right tail. Knowledge that consequences or other outputs are power law distributed increases the confidence in assessing the probabilities associated with events of extreme magnitude. This is unlike analyses based on assuming a normal distribution. In other words, the focus of the analysis is changed from looking at of outliers as something of a nuisance to thinking of extremely rare events that are actually part of the power law distribution.

Importantly – and as significant departure from traditional thinking about variability and central tendency – the tail truly wags the dog. The mean and variance of a power law (the exponent is fractal) distribution are not stable under increasing the sample size – a major and important difference from classical distribution theory based on asymptotic behavior. However, the fractal dimension is asymptotically consistent (statistical) estimator of the population parameter[3] whence a sample was taken. On the other hand, the *scale free* aspect of power laws may be questionable for complex models (Avnir et al. 1998). However, Holme and Kim (2002) construct a network in which the degree distribution follows a power law but also exhibits *clustering*: in some fraction of cases (p), a new node connects to a random selection of the neighbors of the node to which it last connected. Regardless of scalability, uncertainty about the position of the curve requires expert assessment. In the context of rare events, this may have to be done via simulations, perhaps combined with experts' elicitation methods. Second-order uncertainty – the distributions of the shape and location parameters of a probability distribution function – allows a more complete analysis of the uncertainty affecting the distributions themselves and their sensitivity to changes in these parameter's value. Attempting to represent uncertainty in terms of only an output distribution function may be insufficient to support quantitative policy and regulation catastrophic events. In principle, policy should be transparent to the stakeholders and thus mechanisms generating critical outputs cannot be assumed to be generated by system or network that is a *black box*. In other words, it is also important to study the distribution of the inputs and the outputs of a model that is opaque to all but the developer. Correct expert advice requires that assumptions, rules, and results be made crystal clear to all stakeholders.

Survivorship curves (the distributions of the survivals, rather the deaths, from a preventive action and without that preventive action) can also be used to study the effects of exposure on the probability of response. Any combination of probability and magnitude that falls above the curve of tolerable risk is, by definition, not tolerable. In some situations, however, specific combinations of probability and magnitude may be tolerated on grounds other than simply magnitude and probability. Accidents in petrochemical facilities can be compared with accidents in fertilizer plants using summaries such as the FM curves, provided that the same consequences, for example, prompt deaths, are being compared. Each FM curve is parameterized (although there are non-parametric statistical methods that also serve the purpose of data fitting) using data. The process, *estimation* of parameters, is statistical, as

---

[3]A fractal dimension is not integer, unlike a *3*-D object such as a cube. A simple fractal dimension is the quantity of self-similar objects, $D = -(log\ N/log\ s)$, in which D is the dimension that characterizes the invariant relationship between size and number of self-similar objects, and s scales the objects to the entirety of the objects. For example, if $s = 0.25$ and $N = 4$, then $D = 1$ (a line has *1* dimension, it is *1-D*), as we would expect. We can have the common dimensions as well as the uncommon, or fractional, ones. A fractal curve will have a fractal dimension between *1* and *2*, and the larger the dimension (the closer the value tends to *2*) the more the fractal curve will tend to fill the plane.

Table 1.5 Example of statistical issues encountered in statistical model and choice of the model

Issues	Possible solutions
Model form selection	Akaike information criterion, AIC, and Kullback-Leibler, KL, information metric; non-parametric methods; Bayesian model averaging (BMA); model cross-validation; sensitivity analysis for model forms
Variable selection for exposures, responses, confounders, covariates	BMA, cross-validation, bootstrap resampling, classification trees with "bagging". this is a form of resampling that reduces sensitivity of selected variables to random variations in data
Omitted explanatory variables	Test for latent variables. Use discrete mixture distribution models to estimate effects of unobserved variables.
Variable coding	Automatic algorithms, classification trees, MCMC-based adaptive random partitioning to set boundaries. Do not discretize continuous variables
Missing data (missing at random)	Multiple imputations, sensitivity analysis, EM algorithm, data augmentation, and Bayesian modeling
Measurement errors in explanatory variables	Sensitivity analyses, bias correction algorithms, simulation-based corrections (SIMEX), EM algorithm, data augmentation, Bayesian modeling
Exposure metric selection	Use multivariate exposure descriptions rather than any single metric
Multiple testing and multiple comparisons	Adjust p-values of individual tests (e.g., Bonferroni's correction)

shown by the *Anscombe quartet*, fully discussed in Chap. 5, *Uncertain Catastrophic Events: Probabilistic and Statistical Aspects*. Estimating the parameters of most statistical model associated with exposure-response generates several statistical issues, some of which we summarize in Table 1.5.

1.3.2 A Procedural View

In situations where the mathematical expectation is an appropriate criterion for selecting between choices (i.e., optimizing choices that minimize the expected damage from routine events), that choice is either superior or at least as good as a decision based on alternative rules. Granger-Morgan and Henrion (1990) suggest a minimal set of steps for model development. We develop a set of steps that can be used in the statistical analysis of catastrophic outcomes by means of a generalized damage function between exposure (either instantaneous or delayed) and response (either immediate or delayed).

1. *Choice of population* – For any exposure-response relation to be modeled: i) obtain a sample (e.g., random, non-random); ii) the number exposed and, iii) the number of not exposed, and so on. Develop from the theoretical and empirical literature, the risk factors (independent variables), the response variable, the

protocols and coding schemes to be used to recording the data, and address issues associated with missing, outliers, truncated data. There may be more than one dependent variable, as is the case when an independent variable is also a dependent variable in a system of equations.

2. *Select the statistical causal model forms* to be used in modeling the relationships between the variables, testing hypotheses, forecast outcomes, and so on.

3. *Estimate the parameters of the model(s)* in Step 2 from the data obtained in Step 1. This step fits the model form to the data. The estimated values of each model parameter, with the model form based on mechanistic knowledge, determine the probability of failure (the damage itself), given exposure to a hazard. Exposure can be continuous, peak, interrupted, seasonal, and so on.

4. *Assess the model.* This step answers questions such as: Are its predictions valid and accurate? If so, to what degree? Model evaluation implies quantifying and explaining the findings from estimation using statistical tests and develop model diagnostics that indicate how well a model comports with the mathematical assumptions on which it is based. This is a demanding step: it means that the analyst has to conduct analyses and report results from a number of possible methods. For instance, these methods include model cross-validation, Bayesian model averaging, tests of alternative model forms, and so on. The results from this step feed the results back to Step 2 to assure that selected model is appropriate. If necessary, return to Steps 1–3 to refine the model. If this is done, then use appropriate modeling techniques to correct for model-selection and over-fitting biases.

5. *Interpret the results* of modeling in terms of its implications for selecting the optimal or otherwise preferable choice (e.g., answer the question: Is there a statistically significant exposure-response relation that is sufficient for policy-making?). Characterize and discuss all remaining uncertainties and their impact on the optimal or preferable choice presented to stakeholders.

If an unexpected event occurs within the available observations, the reason for that event may be due to a different generating mechanism, unrelated to the data observed. For example, time series of data are realizations from inputs from possibly complex phenomena that may be characterized by transients that cannot be controlled endogenously (by processes such as self-regulation). An example of two transients is the increased deaths in a December 1962 week in London when approximately 12,000 people died from pulmonary disease. This catastrophic event also caused approximately 100,000 hospitalizations. Within that week, the measured *British smog* levels peak clearly preceded (i.e., lagged) the recorded mortality peak. However, when the time series that depicted the temporal evolution of the other risk factors for those winter deaths, and not just the (British) smog levels, are coupled to the time series of those deaths, attribution of the deaths to British smoke become suspect. It has recently been reported (Domonoske 2016) that the causal question regarding the causes of those deaths seems resolved. Summarizing it, the causal agents of the smog principally were sulfate and sulfuric acid particles formed from sulfur dioxide released by coal burning, with the sulfur dioxide changing to sulfuric acid and being inhaled.

1.4 Conclusion

We summarize some of the salient points of the assessment of several types of catastrophic incidents. These have three common characteristics: uncertainty is represented by probabilities and distributions, causation is uncertain, and the consequences are heterogeneous. Although the physics of catastrophic incidents can be complicated, much can be accomplished using simple examples. The critical issue to keep in mind is that small probabilities and large consequences can be combined through their multiplication, but that multiplication results in a number that does not represent the true impact of the catastrophe. The complete representation of the uncertainties and magnitudes should be displayed: these are critical to the correct representation needed by stakeholders and decision-makers. However, as we will discuss in other chapters, some formal analysis can also create critical issues, such as resulting in a choice that is made by a single individual rather than by majority or other rule. Moreover, technical and cognitive biases affect most – if not all – formal analyses. Unfortunately, the effect of those biases may guide an eventual decision.

References

D. Avnir, O. Biham, D. Lidar, O. Malcai, Is the geometry of nature fractal? Science **279**(5347), 39–40 (1998)

H.L.D.S. Cavalcante, M. Oriá, D. Sornette, E. Ott, D.J. Gauthier, Predictability and suppression of extreme events in a chaotic system. Phys. Rev. Lett. **111**(19), 198701 (2013)

R.A. Clewell, C.M. Thompson, H.J. Clewell III, Dose dependence of chemical carcinogenicity: Biological mechanisms, for thresholds and implications for risk assessment. Chem. Biol. Interact. **301**, 112–127 (2019)

C. Domonoske, 23 November 2016, "Research On Chinese Haze Helps Crack Mystery of London's Deadly 1952 Fog". Accessed 6 Sept 2018

M. Granger-Morgan, M. Henrion, M. Small, *Uncertainty: A guide to dealing with uncertainty in quantitative risk and policy analysis* (Cambridge University Press, Cambridge, 1990)

P. Holme, B.J. Kim, Physical Review, https://doi.org/10.1103/PhysRevE.65.026107eview (2002)

IOM, To Err Is Human, Building a safer health system, (Nov, 1999). http://www.nationalacademies.org/hmd/~/media/Files/Report%20Files/1999/To-Err-is-Human/To%20Err%20is%20Human%201999%20%20report%20brief.pdf. Accessed 6 Dec 2019

A. Johansen, D. Sornette, Critical ruptures. Eur. Phys. J. B-Condens. Matter Complex Syst. **18**(1), 163–181 (2000)

A. Johansen, et al, Crashes as critical points. Int. J. Theor. Appl. Finance **3**(02), 219–255 (2000)

B.B. Mandelbrot, P. Pfeifer, O. Biham, Is nature fractal? Science **279**(5352), 783 (1998)

J. Marshall, Yellowstone Supervolcano's Size Exceeds Expectations: Discovery Communications, LLC.; 2012 [updated NOV 27, 2012; cited 2014 JUL 11]. Available from: http://news.discovery.com/earth/weather-extreme-events/yellowstone-park-supervolcano-plume-110414.htm

Medscape, Medical Errors and Outcomes Measures: Where have all autopsies gone? (2019), https://www.medscape.com/viewarticle/408053. Accessed 6 Dec 2019

NAS, *Understanding Risk: Informing Decisions in a Democratic Society* (National Academic Press, Washington, D.C., 1996)

NAS, *Air Quality Management in the United States* (National Academic Press, Washington, D.C., 2004)

NASA, Report to the President, Presidential Commission on the Space shuttle Challenger Accident June 16, 1986. Washington D.C.

P.F. Ricci, S. Tharmalingam, Ionizing radiation epidemiology does not support the LNT model. Chem. Biol. Interact. **301**, 128–140 (2019)

D. Sornette, Predictability of catastrophic events: Material rupture, earthquakes, turbulence, financial crashes, and human birth. Proc. Natl. Acad. Sci. **99**(suppl 1), 2522–2529 (2002)

D. Sornette, *Critical Phenomena in Natural Sciences: Chaos, Fractals, Selforganization and Disorder: Concepts and Tools* (Springer, Berlin Heidelberg, 2006)

M.P. Stumpf, M.A. Porter, Critical truths about power laws. Science **335**(6069), 665–666 (2012)

N.N. Taleb, *The Black Swan: The Impact of the Highly Improbable* (Random House, New York, 2007)

US EPA, *Characterization of Data Variability and Uncertainty: Health Effects Assessment in the Integrated Risk Information System (IRIS), EPA/630–00/005A.* (Environmental Protection Agency, Washington, DC, 2000)

USGS, Questions About Supervolcanoes: USGS; 2015 [updated August 21, 2015; cited 2015 September 25]. Available from: http://volcanoes.usgs.gov/volcanoes/yellowstone/yellowstone_sub_page_49.html

USGS (2018): https://www.usgs.gov/science-explorer-results?es=San+fRancosco+earthquake+probabilities+ Accessed July 8, 2020.

Wikipedia contributors, Black swan theory: Wikipedia, The Free Encyclopedia; 2014 [updated 1 July 2014 05:12 UTC13 July 2014 06:59 UTC]. Available from: http://en.wikipedia.org/w/index.php?title=Black_swan_theory&oldid=615112188

Wikipidia contributors, Volcanic explosivity index: Wikipedia, The Free Encyclopedia; 2015 [updated September 6, 2015; cited September 20 2015]. Available from: https://en.wikipedia.org/w/index.php?title=Volcanic_explosivity_index&oldid=679671652

World Bank, Bangladesh – Arsenic Mitigation Water Supply Project (English). Washington, DC: World Bank (2007). http://documents.worldbank.org/curated/en/309151468002142598/Bangladesh-Arsenic-Mitigation-Water-Supply-Project

W. Willinger, D. Alderson, J.C. Doyle, *Mathematics and the Internet: A Source of Enormous Confusion and Great Potential* (Defense Technical Information Center, Ft. Belvoir, 2009)

Chapter 2
Policy and Legal Aspects of Precautionary Choices

2.1 Introduction

Prospective catastrophic incidents can have a range of adverse consequences by affecting a small area, a region, or nation and often stretch a nation's resource beyond its means. Supranational and national laws, and their secondary legislation, attempt to inform decision-makers about taking precautionary and preventive measures, but they often do so qualitatively. Although administrative regulatory laws deal with concepts (e.g., *fairness*), principles (e.g., *equity*), and procedures (e.g., rules of evidence) and thus the *why* is clear to some, the *how-to* aspect may remain vague. Part of this vagueness is due to the scientific complexities, the prospective outlook of the choices, and the often intergenerational and international impact of those catastrophes. Another part is due to the language used in defining and formalizing causation – from the source of the danger to the consequences that it can cause. Yet another part is the choice of the criterion, from the possible ones, to be used for justifying the selection of one choice over all other alternatives. Finally, incomplete scientific knowledge affects the assessments required by law. In this chapter, we discuss selected elements of scientific analysis of precautionary choices that precede actual decisions and actions by public agencies. Efforts to determine the *how-to* must deal with the effect of partial knowledge and assumptions on the accuracy of analysis of the alternative choices that may be open to society to deal with catastrophic incidents. Assessing prospective catastrophic consequences requires defining their probable impacts and includes unintended consequences, although evident after the fact, that should be anticipated. The causes (or risk factors) are multidimensional; the consequences may range from human deaths to loss of habitat for one or more species. Each consequence has different spatial and temporal characteristics. However, they await us in the future.

Public decisions to prevent natural or human-made catastrophic incidents create a dilemma. Should society wait until scientific uncertainties are resolved to trigger a decision ahead of the catastrophic incident – e.g., causation is clear and

© Springer Nature Switzerland AG 2020
P. F. Ricci, *Analysis of Catastrophes and Their Public Health Consequences*,
https://doi.org/10.1007/978-3-030-48066-0_2

convincing – or should it act even when the uncertainties about the occurrence of the event and its consequences are large but far into the future? Answering this question requires formal analysis to define what is sufficient evidence. That is, what is the best available science, at the time of the analysis, such that the evidence is indeed clear, convincing, and can inform a prospective precautionary decision and when to implement it. These issues are as much scientific as they are legal. Their overall physical context is summarized by the temporal and spatial domains of the *uncertain, serious, or irreversible damages* that can be caused by prospective catastrophic incidents. Temporally, the consequences may be predicted, on the average, to occur after many more generations from the time of their analysis. The costs of protection may be paid in a much briefer period of time, often well within the first generation (it is also likely to be affected regardless of the average return period used in the analysis). These costs are certain: they have to be paid before the catastrophe materializes and are computable through established financial, economic, and engineering methods (Milieu 2011). Moreover, causation is not monolithic: the law limits causation in ways that science does not (Ricci and Molton 1981), and this difference can create difficulties regarding cascading disasters and the attribution of culpability after a catastrophic incident.

This chapter deals with the legally binding commands of the EU's Precautionary Principle (PP) and the US environmental, health, and safety principles that are similar in scope to the PP. All require formally assessing cause and effect because, doing otherwise raises criticisms such as lack of due substantive and procedural process or irrational (acausal) decisions. As we will further discuss in Chap. 6, *Preferences, Time and Dominance: An overview for public choices*, the analysis of available choices to inform public decision-makers, including *do nothing*, ranks the alternatives according to formal criteria, some of which are probabilistic. Each choice should be flexible, in preference to binary choices such as *all or nothing*. Flexibility itself implies that some costs will be larger, in the initial phase of implementation of a decision, to include unexpected changes in the magnitude of the predicted effects. In other words, flexibility accounts for aspects of unpleasant surprises by initially increasing the robustness of the solution, relative to what would be required under normal (e.g., average) design conditions. The methods to assess those choices should be such that they can include changing scientific evidence over time. After the assessment, the results inform decision-makers (e.g., legislators) or their agents (e.g., administrators). The legal-regulatory science-policy process can be summarized in two sequential processes and a main feedback loop:

1. Constitutional or Statutory PP → Procedural and substantive details from administrative or regulatory laws → Identification of perceived or possible catastrophe → Budget allocations for conducting the analyses → Identification of a plausible set of alternative precautionary choices ex ante of the catastrophe → Scientific and technical analysis (identification of probable choices, working hypotheses, assumptions, data, theories, deterministic and probabilistic models, predictions and forecasts, sensitivity and robustness analysis, corrections, final model).

2. Definition of preferable or optimal choice out of the alternative choices analyzed→ Socio-economic analysis → Rank choices using one or more formal criterion of choice → Inform stakeholders of the scientifically-determined preferable choice or choices → Action by regulatory agency → **IF** judicial challenges occurs and is lost **THEN**→ **[FEED-BACK LOOP** to Identify a new plausible set of alternative precautionary choices → 1) → New analyses → New preferable or optimal action].

Before discussing some of the main legal forms of the PP, we repeat the United Nations (UN) Conference on Environment and Development (1992) definition of disaster that may be legally binding: "...serious disruption of the functioning of a community or a society involving widespread human, material, economic or environmental losses and impacts, which exceeds the ability of the affected community or society to cope using its own resources."

This definition explicitly suggests that disasters have international law aspect, in part because of mutual support agreements between two or more nations. Briefly, international disaster law consists of principles, treaties, and case law. For example, the signatories to the UN Hyogo Framework for Action (2007) agreed to reduce disaster risks through setting national strategic goals for risk reduction through regulatory and development decisions by means of a multisector approach. Poverty, land use, and education are critical elements. According to Verchick (2016), from the time of the Hyogo Framework, the number of people in Asia who have been directly affected by natural disasters has fallen by almost one billion. In 2015, the UN adopted the *Sendai Framework for Disaster Risk Reduction 2015–2030*. It concerns building the technical resources to deal with global and regional catastrophic incidents. International legal frameworks include the Cancun Framework (2010) and the Warsaw International Mechanism for Loss and Damage (2013) that address *loss and damage associated with impacts of climate change, including extreme events and slow onset events, in developing countries that are particularly vulnerable to the adverse effects of climate change*. This *mechanism* consists of enhanced *risk management approaches* and coordination between stakeholders regarding finance, technology, and capacity-building (Verchick 2016). Farber (2018) developed a set of US-centered policy *points* that direct the application of precautionary regulations to catastrophic incidents. These apply regardless of the nature of the catastrophic incident. Farber (2018) suggested ten policy points that range from capacity-building, having to account for the disproportionate effects on those at risk, to communications via the media and the impact of catastrophes on the communication and overall infrastructure of the areas affected.

Catastrophic incidents can be extremely dangerous or deadly for particularly fragile segments of the population at risk. For example, heat waves can be particularly dangerous to the more sensitive members (e.g., elderly people) of the general population. In the 1995 Chicago heat wave, almost three quarters of the victims were over 65 years of age. Hurricane Katrina showed that increased risk of death may result from lack of medicines or access to them. Weaker individuals succumb to their preexisting chronic disease, e.g., CVD, diabetes, and respiratory diseases.

The very magnitude of the catastrophic consequences can be grossly underestimated. For instance, Hurricane Maria was first reported to have caused approximately 64 deaths in Puerto Rico. However, the death toll was much higher, *1085, rather than the official 64* (Alexis R. Santos-Lozada, the conversation (January 3, 2018, 6:54 AM), http://theconversation.com/why-puerto-ricos-death-toll-from-hurricane-maria-is-so-much-higher-than-officials-thought-89349). A reason for this difference in the number of deaths is that "[i]n Puerto Rico, deaths are recorded using international classifications that do not capture all of the circumstances surrounding a death that happens following a natural disaster." Specifically, deaths from indirect circumstances, such as *having a cell-phone tower unable to function because of the Hurricane and thereby preventing a 911 call from going through to save the person suffering from a heart attack*, would be classified as resulting from cardiovascular events and not attributed to the hurricane.

There is no doubt that precaution is one of the major driving forces for developing analyses of the choices that society could prefer when facing catastrophic incidents, regardless of their type and nature. At the same time, what precaution means should be discussed because it involves personal, collective, and societal judgments about what to do that can involve costs that range from the trivial to being impossible to meet. We next discuss a legal perspective about what precaution may be in the overall context of rare events likely to result in adverse consequences.

2.2 Legally Binding Precautionary Principles

An early version of legal precaution is Justinian's AD (or CE) 527: *the maxims of the law are to live honestly, to cause no harm unto others, and to give everyone their due*. Causation is essential to Justinian's maxim as it is to various forms of precautionary principles, such as the German *well-founded suspicion*. It was formulated in a lawsuit involving the tranquillizer Contergan® that contained thalidomide (Landesgericht (LG) Aachen, 18, 12, 1970 4 KMs 1/68, 15–115/67, *Juristische Zeitung*, 516). The manufacturer was held to be responsible according to these legal principles:

1. Causal association between risk and source of danger required for intervention does not need to be certain, but must be based on *well-founded suspicion*.
2. Causal association defines tolerance thresholds.
3. The imbalance between drug's risks and benefits suggests that the greater the imbalance the lower the probability of causal association.
4. Tolerance thresholds depend on the state-of-the-art of medical knowledge.

These became part of early (1970s) European Community (EC) (now European Union (EU)) secondary laws (e.g., EC Directives 75/318/EEC, 75/319/EEC). Today, conventions, treaties, Constitutions, federal and national statutes, courts, non-government agencies, researchers, and others define their own forms of precautionary or preventive principles. Although the diversity of definitions may be important, we focus on those forms or enunciations that are legally binding. These require formal justification for actions designed to prevent or mitigate catastrophic consequences and have a common basis. It consists of explicit recognition that: (1) it is impossible to be certain about how, where, and when future catastrophic events can occur; (2) those events are probabilistic; (3) decisions – given scarce and finite resources – should add to social welfare; and (4) uncertainty should not be an excuse to postpone actions.

We briefly discuss other examples of legal[1] definitions and uses of the PP from supra-national (e.g., international agreements such as the Montreal Protocol, the EU's TFEU) to national (e.g., the US Clean Air Act).

United Nations (UN) – Montreal Protocol on Substances that Deplete the Ozone Layer (Preamble, Para. 6), (Amended, 1990): … "protect the ozone layer by taking precautionary measures to control equitably total global emissions of substances that deplete it, … taking into account technical and economic considerations and bearing in mind the developmental needs of developing countries."

European Union (EU) – The European Union Precautionary Principle is part of the European Union's Consolidated Treaties – the Treaty on the Functioning of the European Union (TFEU, Lisbon Treaty – Article 191, paragraph 2) – regarding the protection of the *environment* and *public health*, as well as *the prudent and rational utilization of natural resources*. It states that: "Union policy on the environment shall aim at a high level of protection…. It shall be based on the precautionary principle and on the principles that preventive action should be taken, … and that the polluter should pay."

Article 191.3 requires using scientific and technical data and probabilistic (termed *potential*) CBA and includes the socio-economic development of the EU and its regions. The PP has been held to apply to health and safety (e.g., Case T-74/00, *Artegodan GmbH* v. *Comm'n*, 2002 E.C.R. II-04945). Secondary EU laws, such as *Directives* (e.g., REACH) and *Regulations*, apply the PP through complex regulatory processes. The EU Commission (EC) in COM (2000, 2011) has provided the procedural aspects of the PP. It asserts that the PP is a risk analysis and management function invoked when there are probable serious hazards, which require careful scientific analysis and assessment of the uncertainties.

United States (US) – Federal law contains variants of precautionary expressions that range from justifications based from RA to RCBA, defined below. We briefly describe some of the better known. Ricci and Molton (1981) have reviewed key aspects of risk-based statutes and case law for the US.

The Clean Air Act (CAA) §112(f)) requires an *ample margin of safety* to protect public health or prevent adverse environmental effects.

The Toxic Substances Control Act (TSCA) §2(b)(3)) states that single chemical substances and mixtures should not present an *unreasonable risk* of injury to health or the environment;

---

[1] For example, the US Congress must ratify international treaties before they become part of US federal laws; other international jurisdictions may have more direct procedures (Iverson and Perring-2009).

The Federal Clean Water Act (CWA) §405(d)(2)(D) commands that water quality standards should be ... *adequate to protect public health and the environ-ment from any reasonably anticipated adverse effects.* The CWA (§301(m)(2)) states that the EPA has to account for limited causal knowledge about the relation-ship between pollution and health.

In the US, federal law often demands that precautionary actions should be based on risk-cost-benefit analysis (RCBA). In the US, the US EPA (EPA/100/R-14/001, April 2014) justifications for regulatory decisions should include:

- *"Economically meaningful human health endpoints*: ... risks can be monetized using economic valuation methods. This may include additional or different out-comes than otherwise would be modeled in a health risk assessment. ...
- *Expected or central estimates of risk for a given population*: Depending on the context ... it may rely ... on more conservative or upper-bound estimates of risk. ... recognizing that conservative estimates of risk might differ significantly from central tendency and might lead to biased estimates of benefits.
- *A full probabilistic distribution of risk estimates*: Not only does this ... contribute to a better understanding of potential outcomes, but also it enables ... to incorpo-rate risk assessment uncertainty into a broader analysis of uncertainty."

Legally binding precautionary decisions generally state that uncertain causation should not prevent precautionary action. Yet, context-specific thresholds must sepa-rate action from inaction. Moreover, because of the combinatorial explosion of con-texts and uncertainties, an omnibus threshold is implausible. For example, in the context of environmental and occupational health regulation to prevent catastrophic numbers of casualties from human-made production of goods and services, causa-tion is developed by aggregating scientific evidence. The objective is to develop a causal argument that include rebuttals on the validity (probative value) of the scien-tific evidence and the cause-effect adduced by proponents and disputed by oppo-nents. Regulatory law concerning routine, non-routine, and rare events must balance multiple issues. These include: (i) setting levels of acceptable (or otherwise tolera-ble) probability of death or injury; (ii) the choice of scientific models of prospective cause and effect; (iii) defaulting to a conjectural model for rulemaking; and (iv) allow alternative causal models that differ from the default (generally a linear at low dose and no threshold model) if those can be justified. However, few of those regu-lated take on the validity of the agency's default causal model because of cots and difficulty of proving a negative. The agency sets a numerical level of cancer risk and associates it with an acceptable exposure or dose. For toxicants, the general concept is not probabilistic. It consists of the ratio of actual exposure to a known (scientifi-cally determined) acceptable exposure such that the acceptable daily intake (ADI) is less than 1.0.

2.2.1 Probabilistic Uncertainty Affects Precautionary Decisions

A key justification for using probabilities is that causal legal reasoning includes probabilistic analysis (e.g., Judge Learned Hand' test in *US* v. *Carroll Towing*, 159 F.2d 169 (2d Cir. 1947)). Moreover, the judicial has begun to rely on Bayesian analysis and Monte Carlo simulations. For example, (Fenton et al. 2016) remark that in US state law, in: "*Marks* v. *Stinson* (1994) … the Appeal Court judge relied heavily on the testimony of three statistical experts – one of whom used an explicitly Bayesian argument to compute the posterior probability that Marks had won the election based on a range of prior assumptions."

They also point a pessimistic view for the future of Bayesian methods in the judicial systems remarking that: "… the role of probability – and Bayes in particular – was dealt another devastating and surprising blow in a 2013 UK Appeal Court … ruling (*Nulty & Ors* v. *Milton Keynes Borough Council*, 2013) … The Appeal Court rejected this approach, effectively arguing against the entire Bayesian approach to measuring uncertainty by asserting … that there was no such thing as probability for an event that has already happened but whose outcome is unknown."

The key policy issue with Bayesian analyses is that judges seem unwilling to accept the Bayesian *prior* probability or distribution as sound science. This skepticism may be understandable because the form of the prior may be such that it conflicts with the certainty needed with rendering legal opinions – an ex post decision that can be compared with the likelihood used in Bayesian analysis. The prior distribution, used as the subjective distribution that describes the expert's belief based on past experiences and learning, appears to be the Achilles' heel of Bayesian reasoning. On the other hand, the regulatory approaches used by federal agencies such as OSHA and EPA are probabilistic and Bayesian. The FDA has allowed Bayesian methods in other contexts (e.g., medical devices). Regulatory law establishes standards, e.g., time averaged acceptable exposure based on epidemiological and long-term animal bioassays data. Dose-response models, applied to those data, are probabilistic. For example, environmental and occupational standards for determining acceptably low exposures or doses are based on them.

Regarding precautionary thresholds, the UK uses, in the context of civilian nuclear power generation of electricity, the ALARP (As Low as Reasonably Possible) principle. It is a quantitative threshold beyond which available public (and scarce) resources could be more effectively spent on minimizing other consequences. In the US, Title 10, Section 20.1003, *Code of Federal Regulations (10 CFR 20.1003)*, the ALARA, *as low as (is) reasonably achievable*, principle means: "making every reasonable effort to maintain exposures to ionizing radiation as far below the dose limits as practical, consistent with the purpose for which the licensed activity is undertaken, taking into account the state of technology, the economics of improvements in relation to state of technology, the economics of improvements in relation to benefits to the public health and safety, and other societal and socioeconomic considerations.…"

There are other policy criteria, including *de minimis* and *de manifestis*, which are used to justify thresholds below which societal concerns should be either redirected toward other hazards. Moreover, rather than a single threshold, there can be two bounds to form an interval where a regulator may opt to wait for additional information to decide. In this case, the value of information analysis (VOI) can be invaluable as it clarifies whether addition information is worth both the wait and the cost of obtaining it. There has been much discussion of the PP and how it should be formulated (e.g., O'Riordan and Cameron 1994; Sunstein 2002, 2007; Stokes 2008; Sheng et al. 2015; Morodi 2016; Ahteensuu 2007; Peterson 2006).

2.3 Legal Aspects of Precaution: EU and US

Statutory and regulatory actions can be reviewed by the courts: these are the final arbiters of the form and content of the regulation that enable actions to be taken to minimize the consequences of catastrophic incidents. We emphasize that catastrophic incidents can be sudden but can also remain undiscovered (as was the case for thalidomide administered to mothers to prevent nausea when pregnant but causing their children to be born with severe malformations). Although the time delay between exposure and the discovery of the consequences is a critical element for precautionary actions, we can discuss their broad aspects independently of the time of occurrence to understand some of the legal implications of precautionary actions.

2.3.1 Case Law, EU

In this section, we discuss some of the salient aspects of EU case law regarding how the EU judicial addresses issues related to assessing catastrophic incidents after the fact. The European Court of Justice (ECJ) (now Court of Justice of the European Union, CJEU) upheld judicial deference toward the *intent* of written legislation. For example, in *ex parte Fedesa*, the ECJ stated that the standard of judicial review for EC's decisions is the *manifest error or misuse of powers*: a strong test. However, the ECJ also held that policy decisions should be scientifically informed, rational, and based on the Treaties' principles of proportionality, non-discrimination, and legal certainty. Regarding the strength of the scientific evidence, the EU's judicial uses several standards. For instance, *Fornasar*'s strength of the evidence was weaker than that used in the British bovine spongiform encephalopathy (BSE) case (C-180/96, *UK* v. *Comm'n*, 1998 E.C.R. I-2265). In another case, *Cockle Fishers* (Case C-127/02, 2004 E.C.R. I-07405), the ECJ held that a Member State of the EU did not have to consider cost benefit analysis (CBA) when applying the PP to justify its actions. The EU Court of First Instance (CFI) (now General Court), in case T-229/04, *Sweden* v. *EU Commission*, delved into the scientific record of the health effects of paraquat, finding that this pesticide was dangerous enough to trigger the

EU's PP. In *Pfizer* (*Pfizer Animal Health* v. *Council*, T-13/99), it was held that human health takes precedence over economic welfare. The *Pfizer* court held that regulatory agents could rely on the PP to revoke authorizations for antibiotics growth promoters in animal feed even *though, owing to existing scientific uncertainty, the reality and the seriousness of the risks to human health associated with that use were not yet fully apparent.*

To exemplify the way the PP is used in EU legal instruments such as Directives and Regulations, consider Regulation 178/2002, Article 6. It states, inter alia, that *risk assessment shall be based on the available scientific evidence and undertaken in an independent, objective and transparent manner.* Under the EU Treaties, precautionary choices by a member state (MS) must be proportional to costs and no more restrictive than required to achieve the *high level of health protection.* In those analyses, the MS should account for technical and economic feasibility. Accounting for the magnitude of the risk is based on the timely and systematic review of the actions taken conditioned on the ...*nature of the risk to life or health ... the type of scientific information needed to clarify the scientific uncertainty and to conduct a more comprehensive risk assessment.* The ECJ has held that the MSs can select their own level of protection, which is conditional on the state of the scientific knowledge (Cases: 174/82 *Sandoz* [1983]; C-42/90 *Bellon*; *Commission* v. *Denmark*; C-24/00 *Commission* v. *France*). Moreover, the ECJ held that "... discretion relating to the protection of public health is particularly wide where it is shown that uncertainties continue to exist in the current state of scientific research as to certain substances used in the preparation of foodstuffs (Cases: C-24/00 *Commission* v. *Denmark*; C–24/00 *Commission* v. *France*)."

Commission v. *France* also stands for the proposition that, for an authorization to be granted, a *genuine health risk* must be demonstrated by its proponent. This implies that a generalized presumption of a health risk is not sufficient to support a *PP*-based action by an MS, regardless of what its national law states. This is an example of the supremacy of EU law over that of MSs, insofar as the Treaties so permit. The ECJ has generally required that an assertion of risk under the Treaties' *PP* must be factual, as opposed to be a hypothesis. The earlier case *Pfizer Animal Health* (Case T-13/95) held that hypotheses would not be judicially accepted as evidence of future harm. *Pfizer* stands for the proposition that the proponent of a precautionary measure, under the PP, requires factual scientific evidence that raises a *reasonable doubt* about safety (*Solvay* v. *Commission* (T-392/02), *Sweden* v. *Commission* (T-229/04)). *Sweden* dealt, inter alia, with *reasonable doubt* about the probable effect of paraquat exposure. The ECJ found that mathematical models underestimated exposure. It concluded that those results could not be used to permit paraquat use. The ECJ also held that a field study in which a worker was exposed to a level of paraquat 118% greater than the acceptable operator exposure level, AOEL, was good evidence of risk. The product violated the explicit MS's occupational health regulatory prohibition to exceed its AOEL.

Solgar Vitamins et al., v. *Ministere des Finances et de l'Emploi* (C-446/08) held that an MS can act to prevent harm even when an uncertainty threshold regarding the facts (the probability-magnitude of the consequences that triggers action) has not been established. The MS cannot and should not wait for the full burden of the disease to affect society, as stated earlier in *Malagutti-Vezoneht* v. *Commission* (T-177/02). However, *Commission* v. *UK and Northern Ireland* (C-390/07) had required the MS, in setting a *PP*-based measure, to produce *at least a certain amount of evidence* of both damage and causation. Regarding that evidence, under the EU Treaties Art. 129(1), it should be the *best scientific information*. In this context, (C-390/07) deals with uncertainty in both the form of the scientific evidence to demonstrate damage and its temporal aspects. The UK had opted to wait for additional evidence before acting because the damage to known sensitive areas had not occurred. The ECJ held that, because the risk factors were known, the UK should act, although it allowed a 4 years period further to identify the susceptible areas at risk. Additional discussions can be found in Garnett and Parsons (2017).

In *Commission* v. *Spain* (C-88/07), the attempt to set a threshold was administrative. Spain used the *rule-in or rule-out* method to forbid a product because it was not listed in an authorized list of medicinal herbs. In this context, *Afron Chemical* (C-343/09) held that, when risk studies widely conflict, absent analysis by public agencies, they are not legally persuasive. EU's courts generally emphasize *current scientific evidence* (*Cheminova* et al., v. *Commission*, T-326/09). In *Commission* v. *France* (C-333/08), it was held that *a Member State may, under the precautionary principle, take protective measures without having to wait for the reality and the seriousness of the risks posed by the marketing of those products to be fully demonstrated.…* Nonetheless, there are limits to precaution. For example, *[a]n authorization scheme … disproportionate in the absence of prior authorization* can amount to a systematic prohibition. The role of scientific evidence was also at issue: the EC had not assessed risks before taking precautionary action. The CFI (26 (T-13/99) [2002] ECR II-3305) stated that the EU *institutions were not required, for the purpose of taking preventive action, to wait for the adverse effects of the use of the product as a growth promoter to materialize* and *could properly adopt a cautious approach*, despite limited evidence of immediate risk for human health posed by feed additives. In *Alpharma* ((T-70/99) [2002] ECR II-3495), the EC took regulatory action without being advised (as it should have) by the Scientific Committee for Animal Nutrition. The CFI was satisfied that the EC established *exceptional* circumstances accepting that it could make conclusions about risks from the scientific information available to them, even though the correct protocol was not followed. The EU's CFI, in Case C-77/09, *Gowan Comércio Internacional e Serviços Lda* v. *Ministero della Salute*, 2010, held that *precautionary action presupposes the identification of a potential harm and a comprehensive assessment thereof.…* According to *Gowan*, uncertainty can arise because of *insufficiency, inconclusiveness, or imprecision of the results*. Identification of a prospective catastrophe is a long way from asserting that any of the common legal evidentiary standards can be useful to deal with non-existing direct and concrete physical evidence. Furthermore, identifying and discriminating between precautionary versus preventive actions, as done in some jurisdictions, does not affect the characterization of their states of uncertainty. Rather, whatever test is eventually developed, it normatively should combine causal and evidentiary prospective evidence because the state-of knowledge changes over time: the EU's law of evidence equates the TFEU's *high level of protection* with the *no reasonable scientific doubt* remaining.

2.3.2   Case Law, US

The *Ethyl Corp. v. EPA* (541 F.2d 1 (D.C. Cir. 1976) decision by the Court of Appeals for the District of Columbia Circuit (D.C. Cir.) provides an early precautionary norm for the US. This court approved the EPA's rulemaking to reduce tetraethyl lead (an anti-knocking agent) added to gasoline, under the Clean Air Act, CAA, Section 211. *Ethyl* allowed the EPA to regulate under the *significant risk of harm* standard, defined in the CAA, as amended, Sect. 211. The EPA could assume that a significant risk existed at low lead ambient levels of exposure from evidence of dose-response at much higher exposures. In a later case (*Lead Industries Ass'n, Inc.* v. EPA, 647 F.2d 1130, (D.C. Cir., 1980)), the D.C. court of appeals stated that the CAA has a *precautionary and preventive orientation.* This case dealt with the Congressionally mandated *adequate margin of safety* standard, understood as being equivalent to *erring on the side of caution* (42 U.S.C. § 7409(b)(1) (1994)). When federal statutes are unclear about the role of CBA, US courts generally hold that CBA should be used (*Corrosion Proof Fittings* v. *EPA*, 947 F.2d 1201, 1217 (5th Cir. 1991) and *Envtl. Def. Fund* v. *EPA*, 548 F.2d 998, 1012–18 (D.C. Cir. 1976)). However, in *Whitman* v. *Am. Trucking Ass'ns*, 531 U.S. 457, 464–68 (2001), the US Supreme Court held that Section 109 of the CAA precludes CBA in setting health protective National Ambient Air Quality Standards. On the other hand, in *Am. Textile Mfrs. Inst.* v. *Donovan*, 452 U.S. 490, 509 (1981), the US Supreme Court decided that the Occupational Safety and Health Act require *feasibility analysis,* but not CBA.

The 1996 Amendments to the Safe Drinking Water Act requires the EPA to establish drinking water quality standards to protect public health and to use *the best available peer-reviewed science...*, and *data collected by accepted ... or best available methods ...* in risk assessments. The Occupational Safety and Health Act, OSHAct (1970, 29 U.S.C. §651–683, Sect. 6(b)(5)) states that occupational health standards should *... most adequately assures, to the extent feasible, on the basis of the* best available evidence *of no material health impact.* The Toxic Substances Control Act Sect. 26 requires that *... the EPA must use scientific standards and base those decisions on the best available science and on the weight of the scientific evidence.* Section 26(h) lists the qualitative conditions that the EPA should consider in "employing best available science," Regarding *the best available science*, the EPA has developed a set of *science requirements* (82 FR 33727, 2017). These are: "(1) The extent to which the scientific information, technical procedures, measures, methods, protocols, methodologies, or models employed to generate the information are reasonable ...; (2) the extent to which the information is relevant for the Administrator's use in making a decision... ; (3) the degree of clarity and completeness ... employed to generate the information ...; (4) the extent to which the variability and uncertainty in the information, ... are evaluated and characterized; and (5) the extent of independent verification or peer review ... ."

The EPA, under the 1970 Clean Air Act (CAA) Sect. 109, regulates *criteria pollutants* (i.e., CO, SO_2, NO_x, particulate matter (PM), ozone, and lead) and Sect. 112

hazardous pollutants (e.g., carcinogens and mutagens). The standards set by the EPA to protect public health *with an adequate margin of safety*. In *Lead Industries Association, Inc.* v. *Environmental Protection Agency* (concerning a new airborne lead standard), the DC Circuit Court of Appeals agreed with the EPA that *Congress directed the administrator to err on the side of caution....* In *Natural Resources Defense Council, Inc.* v. *Environmental Protection Agency* (824 F.2d 1146 (DC Cir. 1987)) the found that: "… [T]he Administrator's decision does not require a finding that 'safe' means 'risk free' or a finding that the determination is free from uncertainty. Instead, we find only that the Administrator's decision must be based upon an expert judgment with regard to the level of emission that will result in an 'acceptable' risk to health. … Congress, however, recognized in Section 112 that the determination of what is 'safe' will always be marked by scientific uncertainty and thus exhorted the administrator to set emission standards that will provide an 'ample margin' of safety. … Congress authorized and, indeed, required EPA to protect against dangers before their extent is conclusively ascertained. Under the 'ample margin of safety' directive, EPA's standards must protect against incompletely understood dangers to public health and the environment, in addition to well-known risks …."

In principle, federal agencies cannot use conjectures because rulemaking addresses *factually determined* and *reasonable foreseen* conditions. For example, the CRA (5 U.S.C. Sect. 801) extends Congressional review of agency's *rules* by going beyond the APA scope to include agency's guidance and other less formal documents. The CRA requires that some federal agencies report to Congress, which can overturn regulatory actions through a joint resolution of disapproval (through established procedures between Congress and the White House, WH). The WH proposed an EO (in 2018) specifically aimed at the EPA, which states that this "… *proposed regulation is designed to increase transparency of the assumptions underlying dose response models. … The use of default models, without consideration of alternatives or model uncertainty, can obscure the scientific justification for EPA actions.*" Yet, despite its specificity to the US EPA, a connection to other US federal agencies is apparent from the statement that: *[a]lthough not directly applicable to other agencies of the Federal government, it may be of particular interest to entities that conduct research ….*

Regarding regulatory actions, the Administrative Procedure Act (APA) differentiates between *rules* and *orders*. The former concerns future effects and is of general and particular applicability. For instance, *Yesler Terrace Cmty. Council* v. *Cisneros*, 37 F.3d 442 (9th Cir, 1994) held that orders are adjudications; rules are the result of the *rule-making* process. An *order* is a substantive matter regarding *the whole or part of a final disposition… of an agency in a matter other than rule making but including licensing*. An order is specific to individuals and to their dispute; a rule is generic, prospective, and become individualistic after the rule is applied. The Congressional Resolution Act (S.J. Res. 57; signed as P.L. 115–172, 2018) was used for the first time in 2018 to disapprove an agency's *guidance* document because the agency had not followed its administrative process. All other Congressional resolutions dealt with *regulations* produced thorough the APA's process. Although a marketing order's scientific targets aspects are not as well established as those of standard setting through rule-making because they are fundamentally evolutionary, their granting should be procedurally fair and empirically causal, under the FD&CA (21 U.S.C. Sect. 387), the APA (5 U.S.C. Sect. 500), the agency's internal procedures, and case law. The critical statutory aspects are found in the Tobacco Control Act (P.L. 111–31, 123 Stat. 1776) that amends FDC&A by adding Sect. 911 (21 U.S.C. 387 k). The FD&CA Sect. 911(b)(1) empowers the FDA to issue a marketing order provided that *the product is less harmful or presents a lower risk of tobacco related disease*. The FDA reviews the application and appropriateness of the evidence submitted by the applicant before deciding about the order (Sect. 911(a)(d) and (g)). Table 2.1 summarizes some of the salient differences between *rulemaking* and *orders*.

EO 12898 states that: "no policy will result in disproportionally high adverse human health and environmental effects on low income and minority populations compared to the general population in affected communities."

Our work contributes to the use of risk-cost-benefit analysis (RCBA) in the assessment of public choices and tends to ensure that decisions are in the best interest of society as well as accounting for the risks that may disproportionally be borne by the less fortunate segments of society. The Office of Management and Budget (OMB) addressed assumptions (often critical scientific conjectures) made by agencies that use in risk assessment and management methods to either prevent or be precautionary toward prospective adverse outcomes. As an example, the OMB (NAS 2007), underscoring added for emphasis) stated that: "Risk assessments should explain the basis of each critical assumption and those assumptions that affect the key findings of the risk assessment. If the assumption is supported by, or conflicts with, empirical data, that information should be discussed. This should include discussion of *the range of scientific opinions* regarding the *likelihood* of *plausible alternate assumptions* and the direction and magnitude of any resulting changes that *might arise* in the assessment due to *changes in key assumptions*.... Whenever possible, a *quantitative evaluation of reasonable alternative assumptions* should be provided. If an assessment combines multiple assumptions, the basis and rationale for combining the assumptions should be clearly explained."

For very rare events, it is not clear what *reasonable alternative assumptions* might be and how they should be applied. Other agencies have used terms such as *completeness* and *best available science*. These can conflict with the *reasonable alternative assumptions*: the latter can be used to deny (as opposed to falsify) the former. It is not clear how *the range of scientific opinions regarding the likelihood of plausible alternate assumptions* is developed when causal models are incomplete. Is the *likelihood* judgmental, what is it conditioned on, and by whom? The difficulty of dealing with these questions is exacerbated when *critical assumptions* are multidisciplinary. Catastrophic events may benefit from specificity and stratifications that reflect the variety of assumptions underlying their analysis. However, increasing specificity affects the amount of data in a stratum: it may have a single data point. The need for theory and data analysis that affects science-policy suggests

Table 2.1 Key aspects of US federal administrative standard setting by rulemaking and by orders

Federal statutes and other federal law instruments	Key expressions	Comment on the regulatory instrument	Guidance or other
FSP&TCA Sect. 911 g(2)(A)(iii)	*Best available scientific method*	FSP&TCA, Tobacco Control Act,(P.L. 111-31; 123 Stat. 1776) which amends FDC&A by adding Sect. 911 (21 U.S.C. 387 k)	*Given the breadth of the evidence needed to support the issuance of an order under Sec. 911 (FD&CA) ... it is unlikely that a single study ... or a set of studies ... will provide sufficient evidence to support the issuance of an order*
Ibid, Sect. 911 g(2)(A)(iv)	*Reasonably likely reduction*, and *measurable and substantial reduction in morbidity or mortality*		
Ibid, Sect. 911 (l)(1)(A)	Evidence must be adequate		
Order – A substantive matter; an adjudication. The APA (5 U.S.C. Sect. 500) defines it as ... *a final disposition... of an agency in a matter other than rule making but including licensing*	*Deeming rule* for tobacco products defined in the FDCA is extended through the Tobacco Control Act. MTRPs orders concern demonstrating: (i) risk or exposure modification and (ii) other premarket reviews. For MRTPs, the regulatory standard is *public health* based on net risk-cost benefit analysis for the population as a whole	An order is specific to named individuals (such as a corporation), *Yesler Terrace Council v. Cisneros*, 37 F.3d 442 (9th Cir, 1994); internal procedures and case law	The MRTP application is reviewed by the Tobacco Product Scientific Advisory Council. FDA Draft Guidance for Industry, US FDA (June 22, 2009)
Rulemaking may be *legislative*. Other rules may be interpretative policies. *Rulemaking*: APA (5 U.S.C. Sect. 500), Regulatory Flexibility Act, Information Quality Act, Negotiated Rule-Making Act (5 U.SC. Sect. 556	*Types of rulemaking* 1. Informal, notice and comment 2. Formal, trial like 3. Hybrid 4. Direct and final 5. Negotiated	A *rule* is generic and prospective; becomes individual-specific after rulemaking	Notice of proposed rulemaking, published in *Federal Register* (FR). Can combine specific congressional requirements; can be consultative depending on the type of rulemaking used

(continued)

Table 2.1 (continued)

Federal statutes and other federal law instruments	Key expressions	Comment on the regulatory instrument	Guidance or other
US environmental statute: Safe Drinking Water Act 1996 Amendments	*The best available peer-reviewed science...*, and *data collected by accepted ... or best available methods...*	EPA; Rulemaking	EPA should protect public health *with an adequate margin of safety*
US occupational statute: Occupational Health and Safety Act (OSH Act); 29 U.S.C. §651–683, Sect. 6(b)(5)	*On the basis of the best available evidence of no material health impairment*	OSH Administration (OSHA); Rulemaking	OSHA, the agency, has to *most adequately assures, to the extent feasible,* no material health impairment
Judicial review of Agency rules and orders	Generally, the evidentiary threshold standard is the *arbitrary and capricious* choice by an agency	*Judicial opinion*; can be deferential to Agency' interpretation of the wording of a statute	In 2000, the US Supreme Court held that the FDA did not have authority to regulate tobacco products under the FDCA; the new statute that gives authority to the FDA is the FSP&TCA (Tobacco Control Act)
Executive Order (EO)	For example, adds requirements beyond the APA and the relevant federal statute if the issue is *significant*	Judgment can repeal or confirm an Agency's choice	EO 12866 regarding the roles of OMB and OIRA

The FSP&TCA (Tobacco Control Act) deals with *covered* and *finished* tobacco products. Its Sect. 901(a)(1) lists the 93 harmful and potentially harmful constituents (HPAC) of statutory concern

a mixture of admittedly incomplete theoretical and empirical considerations that we summarize as questions:

- *Theoretical I* – e.g., is a biological marker or a public health index a predictor of an adverse health effect?
- *Theoretical II* – e.g., when is a conjectural model (such as interpolation to zero from the responses at the observed data) supplanted by the more recent factual knowledge (e.g., hormesis, as opposed to threshold models or the LNT hypothesis in toxicology and cancer regulations and causal analysis of exposures at low doses)?
- *Probabilistic I* – e.g., what probabilistic conditioning is used in the analysis?
- *Probabilistic II* – e.g., what is the rationale for the choice of a specific distribution or distributions in an assessment? Equally important, how are situations where the theoretical distributions are not known dealt with?

- *Statistical* – e.g., how is heterogeneity defined and analyzed statistically?
- *Mechanistic I* – e.g., how is human error (e.g., in coding a critical sub-routine) assessed?
- *Mechanistic II* – e.g., how is the magnitude of a rare event (in technological risk assessments) excluded from a practical analysis, given that it is an element in the set of known events?
- *Policy I* – e.g., why has new evidence not become part of the regulatory risk assessment?
- *Policy II* – e.g., why rely on factors of safety when distributions (and the percentiles of the distributions) can be used instead? These factors of safety are heuristic and may be combined multiplicatively, unlike engineering factors of safety that are calculated from known properties of materials and then related to the nature of the hazard, e.g., excessive forces.[2]
- *Policy III* – e.g., is a scientific assessment constrained by legal processes?

In any regulatory context, it is also relevant to know when to stop gathering data. To emphasize its importance, Jimmy Savage is reported to have said that he: "… learned the stopping rule principle from Professor Barnard …. Frankly, I then thought it a scandal that anyone in the profession could advance an idea so patently wrong, even as today (1962) I can scarcely believe some people can resist an idea so patently right."

The science-policy context is based on non-technical failures that often occur either jointly or independently. These failures include:

- I, Legislative, administrative, budgetary.
- II, Technological (including common mode failures).
- III, Existential (either false, incorrect, or partially true narratives).
- IV, Imagination, cognition, perception and behavior.
- V, Scientific, (e.g., theoretical conjectures limited by data).
- VI, Legal: willful or negligent behavior by stakeholders, or other actors (foreseen or unforeseen but foreseeable).

The combination of these six types of failures might influence other failures: this eventuality should be examined and discussed. In practice, these failures form a *hexa-failure* that should be considered as a whole. It is often unlikely that a catastrophic incident – event and its consequences – can be imputed to a single type of failure alone. For instance, the *common mode* failure assumes that human error couples with technological failure and incorrect procedures. Yet, some actions to minimize the prospective event of concern may have positive – but unintended – outcomes. An example, financial institutions in New York City had redundant servers centers in New Jersey to deal with the feared Y2K (computer failures) disaster. In the end, the turn of the century was a non-event. The backup facilities that were built in anticipation of Y2K (year 2000) computer bug would crash their systems. Although the Y2K fears did not materialize, the redundancy turned out to be critical after the 9/11 attacks.

---

[2] A rationale may be statistical, but it is not articulated in this proposal, nor is it justified.

The precept is that public expenditures should not occur unless these yield a positive net benefit to society (e.g., EO 12291; EO 12898). EO 12866 (revoking EO 12291) provides some more information by requiring (between other things) that all federal agencies conducts a regulatory analysis in which employment, environment, public health, and safety (as well as interagency actions) must be assessed when the proposed public project can have a yearly impact greater than 100 million dollars. It follows that the critical element of public decision-making is formal analysis. In principle, this analysis should consist of:

- Systematic, reproducible, and correctable ways to deal with multiple dimensions, complex models, and heterogeneous data often characterized by proxies, gaps, and uncertainties.
- Enumeration of all possible acts, states, uncertainties, and consequences (positive, negative, and status quo).
- Decision rules that envelop the ethical (law of ethics) and analytical aspect of any optimal choices, if it exists. Presents alternatives to optimality, e.g., transitive preferences) to inform the decision-maker.
- An accounting of the sequential availability of information including new information formally, for instance, via Bayesian updating.
- Reducing ad hoc choices when faced with analytical difficulties such as having to deal with non-linear multidimensional models and uncertain causation.
- Accounting for optimism, pessimism, and neutrality via alternative decision rules and sensitivity analysis.

2.4   Precaution and Attitudes Toward Risks

The US IOM (2013) addresses the effect of uncertainty on rulemaking, in a noticeable parallel between the EU's view of uncertainty in the PP, using the term *risk* probabilistically. It notes that: "Because some uncertainty analysis has delayed the rulemaking … (there must be some) caution against excessively complex uncertainty analysis … the amount of uncertainty analysis should match the need of the decision-maker."

However, this quotation appears to place the cart before the ox. That is: (i) Why should *excessively complex uncertainty analysis be* avoided? And: How can this avoidance *match the need of the decision-maker*? The correct methods for uncertainty analysis should inform those stakeholders. The methods should be consistent with the *best science* and *best available evidence*, as the IOM (2013) itself states. If the needs of the decision-makers are to know the extent of the effect of uncertainty on the ranking of choices, then these needs cannot be met by any other form of analysis than the correct one – whatever that analysis may be. Although simplifications are appropriate, these should be the result of the correct analysis, rather than driving them. Analysts should normatively be to provide neutral assessments. However, the analysis of uncertainty should capture the upper and lower uncertainty bounds about those neutral values (e.g., the median or the average values, if these are used). Moreover, the cost of a thorough analysis before the fact may be much less expensive than implementing an ill-informed, possibly biased, or even incorrect policy decision. Different individual attitudes toward risk characterize choices and decisions. The best known are risk neutrality and aversion Osimani (2013) which depicts risk and loss aversion using utilities, rather than monetary values (Wakker 2000).[3]

Judicial decisions use qualitative standards concerning the weight of the evidence presented at a trial by the opposing parties. The difficulty is that each evidentiary standard is necessarily probabilistic: the figurative *weight* of the evidence must be included in the closed interval [0, 1]. Yet, those standards are stated qualitatively: for example, the *preponderance of the evidence*. The amount of uncertainty that each standard should contain depends on the standard. However, it is the decision-makers' understanding of the qualitative phrase that determines the actual weight assigned to the evidence being disputed. Because there is no number assigned to each standard, the pivot is 0.50 under the *preponderance* (i.e., the *more likely than not*) standard. Theoretically at least, each party attempts to overcome the evidence from the opposing party by some small probability: e.g., *0.01*. Judge Jack Weinstein, in *US* v. *Fatico, 458 F. Supp. 388 (E.D.N.Y., 1978)*, reports different probability assignments, as subjective degrees of belief, to decide on the overall numerical weight of the evidence according to ten judges of the federal Eastern District of New York court regarding civil, administrative, and criminal standards of proof, such as the *preponderance of the evidence* (i.e., *more likely than not*), and others, column 2, Table 2.2 (Table 5 in *Fatico*).[4] These ten judges' estimates are vague, internally inconsistent, overlap, and spuriously precise. A simple remedy would assign non-overlapping probability intervals, rather than either crisp (e.g., 67%) or vague (e.g., 50%+) numbers.

For us, the question that *Fatico* raises is more general: How can a decision-maker correctly be informed by probabilistic scientific analyses and assess the available evidence of a future catastrophic incident so that it can be used justify a decision to act under the PP or its equivalents? The policies we have discussed are grounded in public health concerns and choices. These are informed by science and balance existing scientific K&I interpreted and aggregated by committee of experts so that the total evidence can inform policy. Aggregation concerns three functions: (i) assessing heterogeneous and uncertain evidence; (ii) combining it; and (iii) developing causal justification for an agency's action. Because the administrative processes for issuing regulatory standards should be procedurally fair and scientifically sound, so should the aggregation of expert opinions.

---

[3] We discuss risk-seeking behavior elsewhere in this book.
[4] Administrative agencies use some of these standards for judging evidence in its totality as well.

Table 2.2 Probabilities assigned by federal judges as reported in US v. Fatico 458 F. Supp. 388 (E.D.N.Y. 1978) concerning four federal rules of evidence

		Probabilities		
Judge	Preponderance	Clear and convincing	Clear, unequivocal, and convincing	Beyond a reasonable doubt
1	50+ %	60–70%	65–75%	80%
2	50+ %	67%	70%	76%
3	50+ %	60%	70%	85%
4	51%	65%	67%	90%
S	50+ %	Standard is elusive and unhelpful		90%
6	50+ %	70+ %	70+ %	85%
7	50+ %	70+ %	80+ %	95%
8	50.1%	75%	75%	85%
9	50+ %	60%	90%	85%
10	51%	Cannot estimate numerically		

2.5 Precautionary Choices: A Frame of Reference

The frame of reference (*FoR*) is expression that restates our conceptualization of the cause and effect. It is predictive and includes the consequences, at time $t + k$, of a choice that minimizes the error between predictions and an objective measure of prevalence, incidence, or magnitude of the consequences. *FoR* is based on experimental research and individual expert opinions to build inputs, outputs, and form causal linkages. *FoR* assumes a prospective reality, contingent on a present understanding of prospective catastrophic consequences, characterized by multiple causal paths. Each arrow accounts for: (i) one or more mechanistic processes consisting of physical and behavioral quantities (e.g., scalars, vectors, and matrices), ranks, and weights; (ii) their dynamics; and (iii) the *interpreters* of these (i.e., an expert's judgment, a report). *Fusion* aggregates (i) to (iii) over all experts:

$$FoR := \text{Prospective Reality}_{t+k} \rightarrow \text{Perceived Reality}_{i,j,t=0}$$
$$\rightarrow \text{Causation[data}\left(\text{observations, experiments}\right) \rightarrow \text{physical and other}$$
$$\text{causal processes} \rightarrow \text{behavioral, public heath modeling} \rightarrow \text{analyses,}$$
$$\text{evaluations, and adjustments]}_{i,j,\,t=1} \rightarrow K \,\&\, I \text{ fusion}_{i=1,\dots,n;\,j=1,\dots,m;\,t+1}$$

The subscript i implies the ith expert input; subscript j implies the jth report from the literature or an ad hoc experiment. The process should be theoretically and empirically justified. It generally is probabilistic, in the Bayesian sense of consisting of priors and likelihoods. *Fusion* integrates probabilistic distributions that represent uncertainty preferentially to crisp probability numbers. *FoR* and its elements are uncertain for at least two reasons. The first is that results are signals of the probable consequences. The second is that experimental data, processes, and analysis are uncertain. *FoR* combines uncertain information and knowledge and requires first propagating and then fusing specific representations of uncertainty. The propagation of probability distribution functions (pdfs) and probability mass functions (pmfs) generally uses Monte Carlo simulations. The latter yields numerical approximations of the distribution of the output, avoiding complicated mathematical formulae and simplifying the analysis.

The PP is based on science-policy considerations that raise important questions. One of them is: What amount of scientific evidence about the causes, size of the consequences, their severity, and the probabilities of the adverse consequences should normatively trigger a precautionary decision? An initial answer can be binary. At one end, decision-makers may choose not to act, opting to wait for the resolution of gaps in knowledge and other uncertainties. At the other, either the immediacy of the threat or its magnitude may force a rapid policy decision, without the scientific facts being sufficiently known.[5] For example, stakeholders may consider *flexible* choices (in terms adaptability to change to prevailing conditions) that combine procedural, planning, staging of resources, and physical protective measures. This solution balances an initial phase of protection that is more costly because it can be increased to support a wide range of protective structures as new knowledge becomes available and compares it with the solution associated with a predicted – but fixed – size of the magnitude of the *design* hazard. The *value of flexibility* can be optimized relative to the more common structural approaches based on a level of protection based on a fixed magnitude and known causation. In this more common case, a safety factor accounts for the design event being itself variable, but with a relatively small percent variability. For instance, suppose that disagreement is anticipated at time $t = 0$, the time the choices are ranked, and are received by stakeholders for further consideration. The value of disagreement is calculated as the discounted value of the net loss from a choice that is the *veto* choice. As time lapses, information is received and compared by independent observers (i.e., an *insider* observes and an *outsider* observer). The comparison, relative to some threshold value likely to trigger action at a net cost $C[\$]$, suggests that the more likely the disagreement, the higher the value of flexibility. Flexibility responds to temporally varying information while accounting for asymmetric opinions; in particular, it allows assessing irreversible choices when there is uncertainty about the consequences. However, flexibility has no value when there will not be future information.

2.5.1   Discussion

Suppose that the legally binding PP for the jurisdiction that uses it is based on RA alone. Assume the hazard is occurrence of a tsunami of an unknown height at the shore. A group of experts assesses it prospectively and presents to the decision-makers three mutually exclusive and fully exhaustive choices:

---

[5] Recollect that, before the fact, the fundamental elements of catastrophic incidents are (i) probable events, (ii) probable consequences, and (iii) mechanistic cause and effect that link the events to the consequences.

- *Less protection*: Assume a 1/100 years wave height with a 5% probability of occurring over the next 10 years; build a 10 meters high seawall which is 200 hundred kilometers long.
- *More protection*: Assume a more infrequent (1/500 years with a 1% chance of occurring over the next 10 years) wave height; build a 30 meters average height seawall (to account for local conditions affect wave height and thus may reduce the height of the wall) that is 400 kilometers long.
- *Do nothing*: Wait for additional information because the state-of-knowledge is insufficient to make numerical predictions.

The question is: How can the preferred *choice* between these three alternatives be selected? That is, given a catastrophic event, a set of choices, and their calculated enumerated consequences, the maximum value of all the minima may be selected. Rawls believes that the maximum of the minima, the deterministic *max-min*, is equitable because (in theory) it guarantees that the well-being of worst-off is maximized. If probabilities are known or can be assessed, as is the case for routine and non-routine events of our work, the minimization of the expected negative consequences is often used.[6] Regrets can be parsed into additive components that correspond to the types of mistakes that policymakers can make (e.g., accept a false positive; a reject a false negative) regarding the equitable distribution of risks Traub et al. 2011). Finally, choices can be assessed using non-probabilistic weights. For example, Hurwicz's *realism* criterion balances individual pessimism and optimism (the individual is both the decision-maker and the analyst) using weights w_i ($0 \leq_i \leq 1$; $\sum_i w_i = 1$). These are subjective values, such that as the w_i increase, the choice becomes more loss averse. We exemplify the max min criterion in Table 2.3, where the *states-of-nature* (states 1, 2, and 3) cannot be controlled; however, choices (1, 2, and 3) are either calculated or determinable and thus can be controlled. The idea is that prospective decisions have controllable and non-controllable aspects which have to be explicitly determined and formulated. Probabilities serve in this task.

Decision criteria reflect individual attitudes and can lead to choices that are not neutral (unlike those that follow the unbiased expected value). For routine catastrophic events, preferential ordering of alternatives can be based on the maximization of the expected value of the net discounted benefits. The reason is that routine events allow averages to be meaningful. Additionally, the *do-nothing* choice, which is always an alternative that may be preferable to its alternatives until the uncertainties, is resolved. This is unlike the PP, as discussed. However, for rare or very rare catastrophic incidents, this criterion is not applicable because the hazard is unique: very large adverse outcomes (e.g., the 1/2500 year tsunami) should not be analyzed by using the expectation, although this value can be calculated (e.g., 1/2500*3,500,000) assuming a constant population for New Zealand of 3,500,000

---

[6] Hersanyi maximizes the expected (average) utility (measured by *utils*, rather than dollars, to account for the decreasing marginal value of money) of each choice and selecting the largest expected value; alternatively; expected disutility is minimized.

Table 2.3 Max-min decision criterion, states, and consequences

	Consequences under choice 1	Consequences under choice 2	Consequences under choice 3
State of nature 1	10	5	20
State of nature 2	5	20	15
State of nature 3	8	6	0
Criterion: Max-min	8	5	0

The optimal choice under the max-min criterion is choice 1, with max-min value 8 (i.e., more safety is better than less)

people. This example illustrates the complications of using probabilistic methods to represent prospective catastrophic incidents and their consequences. Moreover, as shown earlier, a critical difficulty is the difference between probabilities as numbers and their perception. The difficulties are exacerbated by lack of experience (fortunately) with both the magnitude of those losses and their small probabilities. In this case, the question may be: If these difficulties exist, then why bother? There are several reasons for continuing with using probabilistic analyses. One of them is that the alternative to making those calculations is to make policy choices by fiat. Another is that some coherent information is better than none. And, a third is that these analyses provide replicable and modifiable benchmarks: even if incorrect, they become a common ground for subsequent discussions. Finally, more information should be better than less because it provides increasing guidance and can induce additional insights in the analysis. For example, private insurance cannot cover all catastrophic losses and therefore the public sector has had to provide some or all of the coverage. In the US, the National Flood Insurance Program (NFIP) stands out as an example. In administering the NFIP, FEMA sets flood insurance rates on a nationwide basis; however, the premiums may not reflect regional, local, and basin-specific characteristics. FEMA attempts to keep flood insurance premiums sufficiently low so that owners and insurers remain in the NFIP while attempting to avoid risk prone behaviors by individuals or other entities. Additional information would be locality-specific data that would increase the granularity of the actuarial analyses and specialize the premiums to affect local behavior through subsidies or other economic instruments.

2.6   Conclusion

Justinian's ethical command *first do no harm and give everybody their due* has evolved into alternative implementations of the PP to justify prospective and precautionary actions that minimize uncertain adverse consequences from future catastrophes and accounts for their distribution of the adverse consequences over those at risk. We have reviewed legally binding precautionary enunciations focusing on natural and human-made catastrophic incidents. The PP is a Constitutional level command found in Treaties (i.e., in the EU). Its anticipatory and preventive objectives have equivalent principles in US federal environmental, occupational, and safety legislation and are the subject of judicial opinions of both jurisdictions. The commands of the PP, broadly interpreted, can be equitably met if decisions made on behalf of society are correctly analyzed (e.g., using risk analysis (RA) or the more complete risk-cost-benefit analysis (RCBA)). Formal analyses in the context of the PP and most – if not all – legal-regulatory analyses imply causal modeling of uncertain causal factors, linked mechanisms, and the resulting outcomes. These analyses integrate physical, probabilistic, and statistical models to obtain defensible, transparent, and replicable causal – rather than merely associational – results. Although eventual policy decisions may not always be consistent with scientifically preferable or optimal choices, because decision-makers also balance competing political objectives, principled analyses are essential. Ex ante of likely catastrophic events and their consequences, theory holds that rational decision-makers should choose the action that yields the highest expected utility or the least expected adverse consequences. However, when catastrophes are unique events, averaging (namely calculating the expected values) is questionable on grounds other than multiplying very large magnitudes by very small probabilities. We have discussed and exemplified other decision criteria that combine uncertain causation with behavioral assumption regarding stakeholders' attitude toward risks and losses. For example, the criterion of selecting the maximum of the minima, *maxmin*, can be preferred to the maximization of the net monetary expected value.

Acknowledgments I thank H-X Sheng and T. Bachman for comments on earlier drafts of this work.

References

M. Ahteensuu, Defending the precautionary principle against three criticisms. Trames **11**, 366–381 (2007)

D. Farber, Response and recovery after María: lessons for disaster law and policy. Revista Jurídica **743**, 87 (2018)

N. Fenton, M. Neil, D. Berger, Bayes and the Law. Ann. Rev. Stat. Law **3**, 51–77 (2016)

K. Garnett, D.J. Parsons, Multi-case review of the application of the precautionary principle in European Union Law and Case Law. Risk Anal. **37**(3), 502–516 (2017)

IOM (Institute of Medicine), *Environmental Decisions in the Face of Uncertainty* (NAS Press, Washington, D.C., 2013). https://doi.org/10.17226/12568

Milieu Ltd, the T.M.C, *Asser Institute and Pace, Considerations on the Applicability of the Precautionary Principle in the Chemicals Sector* (Milieu Ltd., Brussels, 2011)

Morodi TJ, The precautionary principle and public environmental decision-making in South Africa: an ethical appraisal, PhD Thesis (March 2016) Stellenbosch University SA

NAS, *Scientific review of the proposed risk assessment bulletin from the office of management and budget* (National Academies Press, Washington, D.C., 2007)

T. O'Riordan, J. Cameron (eds.), *Interpreting the Precautionary Principle* (Earthscan, London, 1994), pp. 229–251

B. Osimani, An epistemic analysis of the precautionary principle. Dilemata **11**, 149–163 (2013)

M. Peterson, The precautionary principle is incoherent. Risk Anal. **26**, 595 (2006)

P.F. Ricci, L. Molton, Risk and benefits in environmental law. Science **214**, 1096–1098 (1981)

H.K. Sheng, P.F. Ricci, Q. Fang, Legally binding precautionary and prevention principles: Aspects of epistemic uncertain causation. Environ. Sci. Pol. **54**, 185–198 (2015)

E. Stokes, The EC Courts' contribution to refining the parameters of precaution. J. Risk Res. **11**, 491–507 (2008)

C.R. Sunstein, Probability neglect: Emotions, worst case analysis, and law. Yale Law J. **112**, 61–107 (2002)

C.R. Sunstein, The catastrophic harm precautionary principle. University of Chicago Law School, Issues in legal scholarship (art. 3), (2007), pp. 1–29

S. Traub, C. Seidl, U. Schmidt, M.V. Levati, Friedman, Harsanyi, Rawls, Boulding – or somebody else: an experimental investigation of distributive justice, (University of Kiel, Economics Working Paper, 2011)

UK Government Office for Science, *Blackett Review of High Impact Low Probability Risk* (UKGO, London, 2011)

M.R. Verchick, Disaster law and climate change, in *Climate Change Law*, ed. by Edward E. Publishing, D. A. Farber, M. Peeters, (Edward Elgar Publishing, Cheltenham; Northampton, 2016)

P.P. Wakker, Uncertainty aversion: A discussion of critical issues in health economics. Health Econ. **9**, 261–263 (2000)

Chapter 3
Catastrophes, Disasters, and Calamities: Concepts for Their Assessment

3.1 Introduction

We develop a simplified understanding of cause and effect relationships associated with the consequences generated by future natural and human-made catastrophes as probable catastrophic incidents. Public agencies and experts assess those incidents and inform stakeholders about decisions that may be taken to protect those at risk. The context of each future catastrophic incident defines its probable consequences and the ability of a country prospectively to deal with them – if it can. Sound scientific analyses precede and inform decision-makers and stakeholders and should provide decision-makers with a rational (e.g., axiomatically sound) ranking of choices that inform eventual policy decisions. Inevitably, however, biased perceptions, beliefs, ideology, and political considerations, to mention a few, cloud what should be a neutral – or unbiased – view of an eventual decision. As we will discuss and exemplify, scientific analyses can formally account for different attitudes toward catastrophic events. There are scientific criteria that mathematically represent a decision-maker's pessimisms, neutrality, or optimism toward the available choices. The best or preferable choices under each of these attitudes can be calculated, independently assessed, and reviewed. Legally, science-based assessments of societal risks are based on peer-reviewed evidence, collectively assessed by panels convened by learned societies and, after the final decision, may be adjudicated by the courts. In all instances, assessing prospective disasters requires modeling their cause and effect. As the NAS (2016b) reminds us, *models are everywhere, ..., and any major regulation that is issued by EPA is ultimately based on a model.*

We take a scientific-technical focus that integrates disaster analysis with the elements of precautionary choices as well as prevention through codes, international standards, and so on. We explore how a future reality, conditioned by past events, can be modeled and its adverse effects predicted. Predictions inform stakeholders about probable events, the size of their consequences, their severity, and other attributes of concern to policy science. Our focus is on analysis and evaluation of choices

© Springer Nature Switzerland AG 2020
P. F. Ricci, *Analysis of Catastrophes and Their Public Health Consequences*,
https://doi.org/10.1007/978-3-030-48066-0_3

rather than how to manage disasters ex ante their occurrence, e.g., via prepositioning resources. Although we discuss methods and criteria for determining optimal – or at least preferable – choices, our work is limited to scientific causal aspects of disaster law. Catastrophic incidents fall between the wholly unforeseeable natural events to foreseeable human errors and include consequent (e.g., cascading) events and their consequences. The event-specific view of those incident considers one or more extremes to facilitate assessments. For example, according to the NAS (2016) in the context of climate change, one such extreme is the *occurrence of a weather or climate variable above or below a threshold value near the upper (or lower) ends of the range of observed values of the variable.* This value may be an interval developed from historical, simulated, or both data from a (finite and context-specific) *event space* Ω. This suggests knowing past sets of recorded events or events. An event-space Ω includes unobserved, but physically probable events, regardless of what they actually are: that space should be complete. Causation also has its event space: there are several causal models: Ω_M. Each Ω implies some form of sampling. Gass and Fu (2013) state that a model is ... *an idealized representation—an abstract and simplified description—of a real-world situation that is to be studied and/or analyzed.* Modeling is generated by connecting the needs of decision-makers, established through statutes and regulations, with the scientific skills of developers. Those provide, in different ways, the structure of the representation that is going to be formalized through mathematical or other means. Some models allow the simulation of agents' behaviors under specific threat or other stresses (http://cphss.wustl. edu/Projects/Pages/Modeling%20Retailer%20Densityaspx (accessed Oct. 7, 2018)). Agents are individuals that may form groups with whom there may be primary and other lines of communication. For example, the national centers of excellence for modeling includes the Framework for Reconstructing Epidemiological Dynamics (FRED); the *EpiFire* simulator of the spread of epidemics and natural or malicious infectious disease; and many others, depending on the center of excellence that developed them (https://inscientioveritas.org/midas-nigms-infectious-disease-modeling/ accessed September 7, 2018). Modeling informs science-policy through generating outputs such as the trajectories of consequences, tendencies toward or away from some normal equilibrium, and account for their uncertainties. The NAS (2016c) describes sources of uncertainty regarding model-based predictions or forecasts. We discuss the forms of uncertainty that affects modeling by relating them to the development of damage functions (e.g., study of the effect of toxic and carcinogenic agents on humans). What we list applies to most damage functions (e.g., from diseases in humans to effects on animals, plants, and even inanimate objects such as corroding a very rare marble structures) as follows:

- Description of the mathematical structure that represents the relationship between causes and effects. The analysis should be predicated on research hypotheses that unbiasedly inform policymaking.
- For dynamical models such as those represent the spread of epidemics or the flow of water over time and space, the initial and boundary conditions (where the modeling starts and how it is confined) must be carefully assessed. These

conditions are imposed on the model by modelers. For instance, the value of an independent variable at time $t = 0$ is an initial condition.

- Models themselves are uncertain because they cannot fully capture the more complex reality that they are designed to represent and to predict. This is an overall uncertainty that describes the difference between the model's structure and the reality it reproduces. For example, a linear model approximates (at least locally) a quadratic model; changes in long-term average conditions also imply changes in their extremes.
- Some physical variables cannot be negative (e.g., mass in kilograms) or the mechanisms cannot produce outputs greater than a specific threshold because that mechanism would defy known physical laws. Mathematically, however, a negative value is both computable and correct.
- Probabilistic parametric uncertainty assumes that the distribution function used in modeling correctly represents uncertainty. For example, the sample mean $\bar{x}$ is an estimate of the population parameter, called the population mean, μ, within upper and lower confidence limits about the estimated value: $pr(\bar{x} \pm k*s/\sqrt{n}) = 0.95$. This interval represents the uncertainty about μ. Here n is the sample size, k is the upper critical value for either the z or t distributions, s is the estimated sample standard deviation, and 0.95 is the confidence level selected by the analyst. In this instance, the distribution function is the normal distribution. In general this assumes a probabilistic tendency toward a very large sample even though the sample used is small.
- Probabilistic non-parametric uncertainty: no distribution function is assumed.
- Conditional probabilistic uncertainty which is best understood as the probability that accounts for some specific quantity being known and how that knowledge affects another random variable. For example, the probability of the data $pr(D)$ is different from the probability of the data given a hypothesis $pr(D|H)$, which in turn is different from $pr(H|D)$, the posterior distribution. The quantity being sought from the multiplication of the other two terms, $pr(D)$ and $pr(D|H)$, yields $pr(H|D)$. These three quantities (more generally as probability distributions) form the core of Bayesian analysis.

Because models inform regulatory science policy, it is important to establish how a legal system has addressed modeling efforts because, as we have discussed in Chap. 2, models and their results are part of the evidence introduced for causation. Fisher, Pascual, and Wagner (Fisher et al. 2010, references omitted) find that: "... legal arguments directly relevant to models have included: that the procedures for considering a model were procedurally improper; that a model was oversimplified; that a model did not apply to a specific factual situation; that the assumptions embedded in a model were incorrect; that the data and statistics used in a model were flawed; that the model was not subject to adequate peer review; and that there was a better model or source of information which could be used. There have also been some striking examples of where courts and tribunals have ruled that a specific model was incorrect."

The negative view may be the result of the *battle of the experts*. For example, litigation involving models-based cause, effect, and results requires judges and lawyers to assess the scientific aspects of the models' formalism. Lawyers for the parties to a controversy hire experts that inform juries or judges that also act as fact-finders, often at great cost to their clients. Experts for one party will attempt to contradict the scientific evidence develop by the experts for the other party. Judges are lawyers, not modelers. The fact-finders (e.g., lay jurors sitting as a jury) render judgments based on their assessments of those debates. On appeal, as it often happens, judges review that evidence as a function of precedents and interpret that evidence in the context of statutes, administrative procedure, and rules of evidence. It is in this phase of policy science that causation is the critical scientific and legal hurdle.

3.2 Cause and Effect

A simple version of some of the arguments that affect empirical causation, as will be discussed in this Book, is depicted in Fig. 3.1. In the three depictions in Fig. 3.1, the continuous line – the trajectory – is obtained by integrating a known model form: namely, a differential equation. The trajectories we show are solutions of differential equations with a common initial condition $y(t = 0)$. Hence, there is no fitting to data because the trajectories are mathematical prediction from a known mechanism that is formalized through a differential equation. Figure 3.1 depicts alternative solutions to a dynamical model.

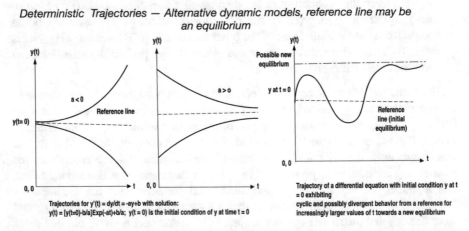

Deterministic Trajectories — Alternative dynamic models, reference line may be an equilibrium

Fig. 3.1 Simple aspects of causation using the solutions of deterministic trajectories of dynamical systems with different trajectories that exhibit departures, convergence, and fluctuations about a reference line that represents an equilibrium state for the system represented by the differential equations. The initial conditions are given as $y(t = 0)$

The implications of a linear equation, Response = a + b*Exposure, can be exemplified by relating exposure to a specific air pollutant PM2.5 to mortality (as number of deaths per 100,000 individuals). The reality being assessed is whether the higher the concentrations of this air pollutant, the higher the mortality rate for those exposed. Asserting, through an equation, e.g., $Y = 3.00 + 0.500*X$, is purely statistical and descriptive: it is an association, out of many other possible ones, with no theoretical content. This class of results can be generated and understood using McLeod (2011) http://demonstrations.wolfram.com/AnscombeQuartet, developed in Chap. 5, *Uncertainty: Probabilistic and statistical aspects*. It is a remarkable example where the same linear model applies to very different data sets that do not exhibit linearity, contain an outlier, or are obviously non-linear. This causal issue is complicated because the diseases associated with these particles are multifactorial. In other words, many other exposures are also associated with the mortality used in the equation. Exposure itself is generated by physical (e.g., combustion) mechanisms, and the PM particles are dispersed by the wind and other physical factors. This requires using dynamical models in which both time and space are used. In other words, the behavior of the plume is seen as a three-dimensional object in the space: concentration, time, and distance. For example, the US EPA (May 2014) *Guidance for PM2.5 Permit Modelling* states (citations omitted) that: "The complexity of secondary $PM_{2.5}$ formation has ... presented significant challenges for the identification and establishment of particular models for assessing the $PM_{2.5}$ impacts ... EPA's judgment in the past has been that it was not technically sound to assign with particularity specific models that must be used ... the appropriate methods for assessing $PM_{2.5}$ impacts are determined as part of the normal consultation process with the appropriate permitting authority."

The choice of a dynamical model that, for instance, predicts the average trajectory of a plume of particles emitted from a stationary source, such as a power plant, can have major policy implications when applied to assessing proposed policies concerning prospective catastrophic outcomes. The EPA (2014) stated that "The assessment of primary $PM_{2.5}$ impacts ... should be consistent with ... specific models as "preferred" for specific types of applications, based on model performance evaluations and other criteria. ... EPA promulgated AERMOD as the Agency's preferred near-field dispersion ... unless another preferred model is more appropriate...." However, having regulatory preferred models does not mean that alternatives to these models are not allowed provided that their scientific basis is sound. The trajectories described in Fig. 3.1 are based on solutions to simple dynamical models – representing mechanisms such as dispersion – using differential equations in which time is the domain. The domain can be spatial and, more completely, combine both time and space suggesting the use of partial differential equations to obtain the appropriate (3-D) solutions. For example, US-wide spatial exposure to hazardous air pollutants such as ozone and fine particulate matter was developed by the US EPA (2015). This agency generated estimates of the annual average ground-level hourly concentrations for ozone and fine particulate matter ($PM_{2.5}$) from 2000 to 2100 under two alternative policy scenarios: global inaction and global action.

3.2.1 Definitions of Disasters, Catastrophes, and Catastrophic Incidents

Science policy discussions of disasters have several concepts, definitions, and principles that often differ depending on their source. We review concepts and definitions from selected influential sources that include the National Academies of Science of the US (NAS) and public agencies such as the US EPA, as well as the EU's Treaties, Regulations, Directives, EU Commission's statements, and others. The IFRCRCS (2002) has suggested that disasters are defined *intuitively* by Rodriguez, Tarantelli And Dynes (200) and as the passage to a state of uncertainty. Perry and Quarantelli's (1998) book *What Is a Disaster* categorizes them from a point of view different from ours. Alexander notes the definitional dilemmas inherent to disasters is due to cognitive differences. He notes that the *Sherman* landslide in Alaska consisted of approximately 29 million cubic meters of earth that slid at about 180 km/hour without killing humans or destroying private property. It was discovered by chance. On the other hand, the *Aberfan* landslide in NSW (Australia), which was approximately 200 times smaller in volume, killed 144 people, of whom 116 were children.

Rodriguez, Tarantelli And Dynes (2007) asks for *a minimum consensus on the defining features ... the characteristics, conditions, and consequences of disasters*. It is necessary to inform about the essential elements within the *cause → effect* path that are common to catastrophes, disasters, and calamities regardless of their source. Such link can be developed at a level of detail that bridges qualitative definitions and legal prescriptions with the formal elements of causation. It does not seem possible unambiguously to define a catastrophic incident with a single number from a specific consequence because each catastrophe has large sets of contexts, perceptions, dimensions, events, consequences, and uncertainties. Hence, asserting that ten or more prompt deaths is a catastrophic incident omits several other essential consequences that, more often than not, are not just prompt deaths. For instance, four deaths and 100 injured can rank a disaster below one that caused 15 deaths and no injuries. For casualties, their continuum is from minor lacerations to death; for environmental consequences, the range is from animals hurt to species extinction. Another consequence may be measured by percent lost habitat. Yet, a catastrophic incident has to be formulated to trigger precautionary analyses leading to ranked choices, and then to a decision. For example, setting a minimum number of human deaths does not define a catastrophe; rather, it does provide that threshold that should trigger public action.

These difficulties result in vague statements. The Robert T. Stafford Disaster Relief and Emergency Assistance Act, *Stafford Act*, (Pub. L. 93–288, as amended, 42 U.S.C. Sect. 5121 et seq.) defines, in Sect. 102, a *major disaster* to mean: "... any natural catastrophe (including hurricane, tornado, storm, high water, wind driven water, tidal wave, tsunami, earthquake, volcanic eruption, landslide, mudslide, snowstorm, or drought), or, regardless of cause, any fire, flood, or explosion, in any part of the United States, which in the determination of the President causes damage of sufficient severity and magnitude to warrant major disaster assistance under this

Act to supplement the efforts and available resources of States, local governments, and disaster relief organizations in alleviating the damage, loss, hardship, or suffering caused thereby."

The World Health Organization (2002), WHO, a UN agency, states that: "[a] disaster is an occurrence disrupting the normal conditions of existence and causing a level of suffering that exceeds the capacity of adjustment of the affected community ... It is the people who matter most, and without the people we have no disaster."

For the US National Emergency Management (6 U.S.C., Ch. 2): "... the term "catastrophic incident" means any natural disaster, act of terrorism, or other man-made disaster that results in extraordinary levels of casualties or damage or disruption severely affecting the population (including mass evacuations), infrastructure, environment, economy, national morale, or government functions in an area."

Other definitions include quantitative thresholds. For example, an early definition suggested *at least 100 people dead or 100 injured or 1 million US $ damage* (Hewitt and Sheehan 1969). More recently, the US Office of Foreign Disaster Assistance, OFDA, (1995) states the definition as: "The threshold for the total number of people killed and injured for inserting them as disasters in the database varies per type of hazard between 25 for earthquakes and volcanoes to 50 for weather related disasters to 100 in man- made disasters."

Regarding K&I, a useful disaster data base is the EM-DAT (CRED). A catastrophe is included in this data if it resulted in at least 10 deaths or affected at least 100 individuals, or there was a declaration of a state of emergency or an appeal for international assistance. Disasters are classified as *significant* if they cause 100 or more deaths per event, damage is 1% or higher of the total annual GNP, the number of the people affected equals or exceed 1% of the total national population. An equally important database *NatCatService*, developed by Munich Re, contains six catastrophe classes (i.e., category 1 through 6) that depend on the magnitude of their monetary or human impact. These range from a natural occurrence with very low economic impact (Category 1) to a "great natural catastrophe" (Category 6) (Munich Re 2006). The latter accords with the definitions used by the United Nations, which defines *disaster* (UN Office for Disaster Risk Reduction, (UNISDR, https://www.unisdr.org/we/inform/terminology accessed August 29, 2018) as: "A serious disruption of the functioning of a community or a society at any scale due to hazardous events interacting with conditions of exposure, vulnerability and capacity, leading to one or more of the following: human, material, economic and environmental losses and impacts. ... The effect may test or exceed the capacity of a community or society to cope using its own resources, and therefore may require assistance from external sources, which could include neighboring jurisdictions, or those at the national or international levels."

Following the Sendai *Framework for Disaster Risk Reduction 2015–2030* (para. 15), the UN's ISDR Office ranks disasters as follows:

- Small-scale disaster: a type of disaster only affecting local communities which require assistance beyond the affected community.
- Large-scale disaster: a type of disaster affecting a society which requires national or international assistance.

- Frequent and infrequent disasters: depend on the probability of occurrence and the return period of a given hazard and its impacts. The impact of frequent disasters could be cumulative, or become chronic for a community or a society.
- A slow-onset disaster is defined as one that emerges gradually over time. Slow-onset disasters could be associated with, e.g., drought, desertification, sea-level rise, epidemic disease.
- A sudden-onset disaster is one triggered by a hazardous event that emerges quickly or unexpectedly. Sudden-onset disasters could be associated with, e.g., earthquake, volcanic eruption, flash flood, chemical explosion, critical infrastructure failure, transport accident.

The United Nations Development Program, UNDP, (2015) defines a disaster as: "... a serious disruption of the functioning of a community or a society causing widespread human, material, economic or environmental losses which exceed the ability of the affected community or society to cope using its own resources'... A natural hazard will only cause a disaster if, (a) a community or societal asset is exposed to it; and, (b) if the societal community/asset is vulnerable. For example, an earthquake in an uninhabited area may have no impact on a society and therefore is not a disaster. Vulnerability describes the degree to which a community is susceptible to a hazard's impact. It is determined by physical, social, economic and environmental factors and processes."

The OECD (2012, https://www.oecd.org/gov/risk/G20disasterriskmanagement.pdf) summarizes some of the key qualitative and quantitative aspects of the analysis of disasters as follows: "[t]he expression of likelihood as a variable to determine risk needs to reflect the type of hazard, the information available and the purpose for which the risk assessment output is to be used. For instance, a return period can be formulated for many hazards as the average length of time in years for an event of given magnitude to be equaled or exceeded. A 7.0 M_w earthquake with a 100-year return period at a given location means that an earthquake of 7.0 M_w, or greater, should occur at that location on the average only once every 100 years."

The distribution of the adverse consequences from catastrophes should be of serious concerns for analysts and stakeholders. After magnitude and probability of occurring, the concern of the analyses should be the distribution of the consequences over groups of individuals who are at larger risk, by being more susceptible to injury or death, than the average. Moreover, natural catastrophes can cause migrations that may stress the resources and capabilities of the nearby countries toward whom the migration is directed, cause geopolitical instability, and may result in communicable diseases. The host countries may not be able to deal with that increased stress and thus a domino effect may take place, depending on that country infrastructure and wealth.

3.3   Hazard, Risk, Likelihood, and Other Concepts

Terms or phrases used in characterizing catastrophic incidents range from *serious, wide-spread losses,* to *expression of likelihood.* These should be unambiguous so that they can be formally understood. For example, *extreme randomness, plausibility, events associated with extreme randomness,* and *return period* (Table 4, OECD

2012) may be useful as a qualitative guide but at least two of them are vague (i.e., *extreme randomness, plausibility*). There are established differences between *probability, plausibility, likelihood,* causal or correlative *conditional probabilities, randomness,* and other terms. Although it may be true that an *occurrence measure* and a *determination* may yield a *likelihood,* we prefer to use these terms as defined in Bayesian analysis. The likelihood is a conditional probability, pr(D|H): the probability of the data D, given the hypothesis H. Of course, each of these quantities can be distribution functions, rather than just probabilities. The return period is a frequency-based calculation that requires an historical record and some assumptions in which the homogeneity of the events over which frequencies are calculated. The return period itself is the average period of time between the same event (e.g., a specific flow rate) re-occurring.

We develop the scientific foundations that should help to understand key aspects of what those definitions would entail to support qualitative policy debates as found in *Disaster Law* or *International Disaster Law* (Farber and Daniel 2018). The (NAS, 2016a) suggests several *terms used in* [its] *proceedings and definitions that reflect their usage in prior Academies reports.* Our use of the term *physical* most broadly means a tangible or intangible effect or consequence that is either directly or indirectly measured and is associated with a causal mechanism. For example, it can be chemical, biological, physical, cognitive, or other (Table 3.1).

We follow Ippoliti and Cellucci (2018), Jaccard and Jacoby (2010), Magnani, Carnielli, and Pizzi (2010), Kuhn (1962), and Lakatos (1978) to list terms that may characterize causal propositions, assumptions, mechanisms, and results. Those terms include:

1. *Data* (used in the singular instead of plural; singular: *datum*) – observations (e.g., ranging from instruments to international epidemiological studies such as) are accessible but can be infrequent or even singular, unique, or exceptional. Raw data implies a direct observation not changed by some intervention beyond the original measurement.

2. *Phenomenon* (plural: *phenomena*) – interpretations of causal events, regularities, irregularities, described by physical processes explained by theory and corroborated by observations. The latter may not be immediately available. Data may be used to generate testable hypotheses when theory is unavailable. Phenomena can become *laws* in the sense of Newton's laws of motion, thermodynamics, and so on.

3. *Paradigm* – A theory that is either substituted or modified by another. It identifies a possible conflict between data and phenomena because observations imply a theoretical understanding of the phenomenon or else would probably not be gathered because of costly and time.

4. *Hypothesis* (plural: *hypotheses*) – A theoretical construct that asserts a neutral position and that of its positive or negative alternatives. It can be simple or compound. Hypotheses are not unique and are generally under-determined because their combination can be large. They are discussed in the chapters that deal with estimation and inference.

Table 3.1 Typical terms (in italics) used in policy analysis of catastrophic incidents and our discussions [NAS, Characterizing Risk in Climate Change Assessments: Proceedings of a workshop, Nat. Acad. Press (2016c)]

Definitions	Comment	Used in this work as:
Consequence: The magnitude of damage that would result from a hazard	The causation implicit in this definition seems to be probabilistic (or stochastic)	The size of the damage is in physical or other appropriate units. Causation is fundamental to the process: analysis of choices → inform stakeholders → decisions may or may not be taken → actions follow (including no action)
Hazard: The cause or causes of the direct and indirect consequences	Some authors include the term *potential* that is ambiguous and not defined. It generally implies a probabilistic or other uncertainty	A hazard is a physical event with adverse consequences. Potential is a form of energy: e.g., potential energy
Impact: Often a change in the magnitude or frequency of physical outcome	A change stated as an impact implies a difference from a datum to some other datum	The temporal and spatial aspects of change are essential to understanding the magnitude of the consequences. An impact is a summary measure such as the expected value of the casualties or their total number. An impact can be negative, neutral, or positive
Probability: The likelihood that hazard occurs within a specified period of time	The likelihood is a conditional probability; it is important to distinguish between unconditional and conditional probabilities	Probabilities are dimensionless measures of uncertainty
Risk: The NAS (2016c) defines it as the *combination of the magnitude of a potential consequence of a hazard or hazards attributable to climate change and the likelihood that the consequence will occur*	It is not clear why a *consequence*, a clearly defined quantity with physical or other units, is (or should be) combined with *potential*	A combination may be calculated as (*potential∗impact*). When so, the combination becomes a dimensional quantity, its dimension is determined not by the likelihood (which is dimensionless) but by the magnitude of the consequences (e.g., number of deaths), namely the *impact*
Vulnerability: The susceptibility to a hazard or the inability to respond to sudden or seasonal variability by the system at risk	Susceptibility, sensitivity and vulnerability should be defined separately as should the variability over time and space	The level of susceptibility should be formalized as should those characteristics associated with other terms
Adaptation: The ability to cope with the effect of forces or other stresses by adjusting behavior prospectively	Adaptation also occurs in other biological or other systems at risk, but their time constants and cause-effect may be quite different from those for human systems	Actions imply an analysis of choices that inform decisions to act or not to act depending on political or other factors. Actions follow choices and decisions: analysis of choices → inform stakeholders →decisions may or may not be taken → actions (including no action)

5. *Inference* – It can be inductive or deductive. In most practical circumstances, we will discuss it is inductive. Statistical inference implies inductive reasoning from a sample of data to the population from which the sample was randomly taken.
6. *Justification* – The criteria or basis for accepting a specific causal construct.
7. *Falsifiability* – It enhances belief in a theory. The results from a theory should be compared with known facts to assess its level of falsifiability.
8. *Confirmation* – If a fact implies a hypothesis, then the belief in that hypothesis increases but cannot be definitive because there can be other hypotheses that are supported by the same belief. Probability theory cannot confirm a hypothesis exhaustively. Simplifying a theory reduces the number of hypotheses and minimizes false results.
9. *Plausibility* – A theory accords with known data and similar theories. All things being equal, a simpler theory is preferable to a more complicated one (e.g., *Occam's razor*).
10. *Scenario* – The *specific, plausible natural or expected course of events connecting cause to consequences*. It is not certain; it is neither statistical, probabilistic, or a formal prediction. In general, a scenario is developed by experts.
11. *Possibility* – Something that *could* happen.
12. *Credibility* – *How* an event could happen.
13. *Relevance* – The ability to inform about a natural or expected specific aspects of events that connect cause to outcome (or effect).

Disasters are often associated with qualifiers such as: *preventable, unforeseen, probable, not credible, possible,* and others. These qualifiers depend on the circumstances and time either preceding or elapsed, relative to the catastrophic event. Some routine and non-routine technological disasters appear to be preventable, before and after the fact. An example of a before-the-fact preventable disaster is the Texas City refinery explosion, in 2005, that killed 15 people, injured 180, and caused financial losses in excess of 1.5 billion dollars. An ex-post assessment found that the hazard could have been prevented had BP not (seemingly knowingly) violated safety-related administrative and technical procedures and processes (CSHIB 2007). A general term that appears to include all of these qualifiers are *scenarios,* which are discussed by the OECD (2012). This agency suggests several qualitative elements that are context specific and necessary to develop a *scenario* describing the cause and effect of a catastrophic incident. Those include: the hazard itself, its spatial and time characteristics, its intensity (strength), duration, and so on.

Example According to the OECD (2012), the *risk matrix* can be used to assess accident scenario should describe *the event clearly and in sufficient detail to provide a precise and consistent basis for the assessment of an event's likelihood and impact.* A typical risk matrix is depicted in Fig. 3.3 and is discussed by the OECD (2012) and by the US Department of Defense, DOD (2017) (Fig. 3.2).

The *risk matrix* is a means to determine *whether the risk is imminent enough to be worth assessing.* The OECD states that "[d]ue to the numerous types of risks …

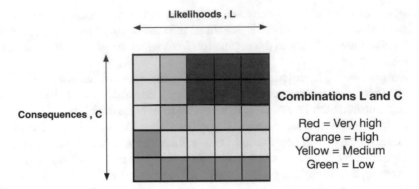

Fig. 3.2 The qualitative elements of a risk matrix: likelihoods, L, and magnitude and type of consequences, C. The SW combinations of L&C (as squares) are the least worrisome; the NE squares are the most worrisome. Each matrix applies to homogeneous consequences and their likelihoods that can be understood as either subjective or frequentistic conditional probabilities). Red = Very high, Orange = High, Yellow = Medium, Green = Low

and the infinite potential risk scenarios, the decision must be made what risks are prima facie important enough to assess. Countries may establish a clear time horizon beyond which a risk scenario is not considered. For example, the event in question might occur within 1 year, 5, 10, 15, 20 years or more. If it has a sufficiently low likelihood of occurring within the next 5 years investment in emergency response capabilities might not be justified in the immediate term. This process helps to prioritize the types of risk scenarios for which investments are needed now in prevention, mitigation or emergency response capabilities to reduce or manage disaster impacts. Different time horizons may be used based on the type of risk assessment performed."

The risk matrix approach can be a *screening* level, quasi-probabilistic, assessment. It helps to decide whether frequencies are sufficiently well understood and the hazard imminent to either warrant further assessment or not. This agency discusses scenarios in terms of *relative likelihood* and *impact*. However, it is not clear what the thresholds for a decision are. As Fig. 3.3 depicts, thresholds are interval valued. If we were to think of a sharp threshold, for example, should the 1/1000-year event be *further* investigated? Analysts can do more detailed analysis to sharpen the confidence in the qualitative results. Nonetheless, what should be done is a policy determination. Risk matrices are theoretically questionable (Cox 2008) in part for the following limitations:

1. Users can correctly and unambiguously compare only a small fraction of randomly selected pairs of hazards.
2. Users can assign identical ratings to quantitatively different risks.
3. Users can assign higher qualitative ratings to quantitatively smaller risks to the point where, with hazards that have negatively correlated frequencies and severities, imply worse-than-random decisions.

Level	Likelihood	Probability of occurrence
5	Near Certainty	> 80% to ≤ 99%
4	Highly Likely	> 60% to ≤ 80%
3	Likely	> 40% to ≤ 60%
2	Low Likelihood	> 20% to ≤ 40%
1	Not Likely	> 1% to ≤ 20%

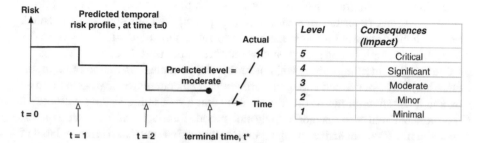

Level	Consequences (Impact)
5	Critical
4	Significant
3	Moderate
2	Minor
1	Minimal

Fig. 3.3 Relationships between the elements of the risk matrix. (Motivated by DOD 2017)

4. The assessment of likelihoods and consequences and resulting risk ratings require subjective interpretation such that different users may obtain different ratings of the same quantitative risks.

The DOD defines risk as being characterized by *probability of occurrence and consequence*; the rules for combining these two quantities are not defined. They may however be implied and yield numerical results using the algebra of probabilities, such as the expected value or other formula. The US DOD defines *risk* as a: "... potential future event or condition that may have a negative effect on achieving program objectives for cost, schedule, and performance. Risks are defined by (1) the probability (greater than 0, less than 1) of an undesired event or condition and (2) the consequences, impact, or severity of the undesired event, were it to occur."

An additional element is the lower left panel curve which depicts *risk burn-out* behavior over time. It implies a decreasing level of concern. We have included an unexpected event (dashed arrow), as might be caused by a sudden change in conditions from a DK event. The interpretations of likelihoods as probabilities and consequences complete the basic structure of the method.

The development of a situation-specific risk matrix is based on agreements between those concerned who generally are experts. A critical issue with asking experts for their opinions is that they can be reticent to go out on a limb (in either direction of the magnitude). Although the magnitude of the catastrophe itself can be very large, it may be understated to meet some preconceived political, ideological, or other position by an agency. In part to deal with these issues, Julian Talbot provides an extended risk matrix analysis, https://www.juliantalbot.com/single-post/2018/07/31/Whats-right-with-risk-matrices, accessed August 30, 2018). In any case, the relations between component ratings and the overall risk rating put strong constraints on rating systems (Cox 2008). For example, if quantitative ratings are used, then conditions *5* and *6* imply that the aggregation formula used to combine component ratings into an overall risk rating must be multiplicative, i.e., the overall risk rating is proportional to a product of its component ratings. Multiplicative aggregation of quantitative ratings satisfies properties *1* through *4*. On the other hand, if only qualitative rankings are used for the components, then it turns out that there is no qualitative ranking system that can assign coherent overall risk rankings (meaning complete, transitive rank-orderings with ties allowed) based on arbitrary component rank-orderings in such a way that principles *1* through *4* are satisfied. Similar limitations hold for aggregating fuzzy ratings of linguistic labels or scales (e.g., *H*, *M*, *L*, and *N*), depending on how they are formalized. In other words, qualitative component ratings alone without further mathematical operation (e.g., multiplication) may not contain enough information to be coherently aggregated into an overall qualitative rating. Another concern is that a risk rating system with only a few possible outcome categories may not produce enough information to assist deciding if it is not inclusive enough. For example, a *3*4* matrix assigning a label of *H*, *M*, *L*, or *N* to each of three components (Hazard, Exposure, and Impact) can provide only a small amount of information (technically, at most six bits of information, equal to the information content of six tosses of a fair coin) to guide decision-makers. Of the much larger quantities of potentially useful and relevant information collected and entered into such a rating scheme (several hundred bits at a conservative estimate), almost all is lost in aggregation during the rating process (Cox 2008). The small fraction that remains (*6* bits in this case, or even less if the probabilities of the *12* cells are not all equal) may be insufficient for effective decision-making, which typically requires at least enough information to discriminate among alternatives that have very differently preferred outcomes. The minimum amount of complexity and information required for a classification system (e.g., hazard rating systems) can be analyzed via techniques from information theory and statistical learning theory. A key insight from such formal analysis is that a classification system that lacks enough complexity to discriminate well among essentially different situations may lead to poor decisions, i.e., ratings with high error rates and high expected losses from decision errors.

The two critical components of risk matrices are described as likelihood and consequences. The latter is generally defined by the size and nature (e.g., prompt deaths, within a 24-hour period of the event causing it). However, likelihoods are often stated phrases, rather than numbers. The MITRE Corporation Systems Engineering Guide (2014) also characterizes probability intervals of occurrence to account for descriptions such as *likely to occur* (i.e., from 61% to 90%) suggesting that *risk rating is based on the probability of impact and the level of impact (manual mapping approach)*.

An even simpler method than risk matrices, the *Italian flag* method, further simplifies screening analyses. This method, according to the UK *Blackett Review* (2011) has three bands: red, white, and green. The width of the bands represents the probability of these three states: success (green), incomplete (white), and failure (red). It is an attempt to introduce, ex ante the event of concern, a state that corrects for the *excluded middle* in binary choices (e.g., head or tail, but not both). The example that the *Blackett Review* provides introduces third element. It changes the sampling frame for the event's probability success or failure to one that includes the landing on the edge of the coin, which the *Blackett Review* labels the *loss of a coin* that was tossed to determine success or failure. The flag only has two colors that – on the average, for an infinite number of trials – would be half red and half green. The true sample space is head, tail, and side, because the physical state-space of the coin consists of these three because the coin is a cylindrical object. Certain coins have their third side semicircular rather than flat. This in practice prevents landing and staying upright on that side.

As a final comment, prospective assessments may use *heuristic* methods. Those are qualitative, subjective rankings through simple numerical score that may be informed by historical evidence or personal experience. Unfortunately, Kahneman and Tversky (1984) established that individuals do not assess prospective events according to probability laws. Rather, the framing of the question and other cognitive biases affect their opinions more so than the probabilities or likelihoods. Moreover, many individuals incorrectly assess probabilistic information and fail to account for the mathematics that should be used to combine probabilities. On the other hand, it has been shown that decision-makers can make *potentially optimal decisions* using heuristics that use little information and process it quickly. Seemingly, people implicitly trade a strategy's cost (e.g., the effort required to use the strategy) against its benefits (e.g., the perceived accuracy of the strategy) and select that is associated with strategy that has the best effort-accuracy trade-off. *Fast-and-frugal* heuristic techniques suggest that individuals bound their rationality using simple, rapid, and frugal rules to support robust and accurate strategies (Gigerenzer and Gaissmaier 2011).

3.4   Conclusion

The discussions suggest that there are hazardous situations in which approximate and often subjective probabilities (termed *likelihoods* in these specific contexts) and consequences that can range from monetary to prompt casualties identify different zones of concern. A consequence falling in one of those zones (or cell of the matrix) may lead to either taking action, wait, or be unconcerned. Rapidly evolving events with almost immediate and serious consequences force rapid analysis and decisions. The combination of magnitude thresholds and likelihoods using matrices is a rough approximation of reality. Its use is different from protracted scientific investigations that often require long periods of time between obtaining scientific results through formal analysis. Although the concern with routine and non-routine adverse events that generate serious consequences for society is common to both, the main difference is the form of the analyses and causal factors leading to it. In the former, causation is two-dimensional – likelihood and magnitude. In the latter, it is based on mechanistic causation. In protracted causal analyses, these three components combined through explicit modeling of how the pairs should theoretically and empirically occur. Both analyses attempt to provide a sufficient basis for analyzing the almost certain high costs of prevention through different levels of explanation. The balancing of the net costs and benefits inherent is informed by either type of analysis.

References

Below, R., Wirtz, A., Guha-Sapir, D. (2009). Disaster Category Classification and Peril Terminology for Operational Purposes Common accord. Centre for Research on the Epidemiology of Disasters (CRED) and Munich Reinsurance Company (Munich RE). Working Paper 264, 20p. Universite' Catholique de Louvain, Brussels, Belgium

L.A.J. Cox, (2008) What's Wrong with Risk Matrices? Risk Analysis, 28, 497–512. https://doi.org/10.1111/j.1539-6924.2008.01030.x

CSHIB, Chemical Safety and Hazard Investigation Board, Final Investigative Report (Mar. 2007)

DOD, Dept. of Defense, Risk, Issue, and Opportunity Management Guide for Defense Acquisition Programs. Office of the Deputy Assistant Secretary of Defense for System engineering, Washington, D.C., (2017)

EM-DAT. The international Disaster databased. Centre for research on the Epidemiology of Disasters – CRED. http://www.emdat.be/glossary/9 accessed June, 8, 2019

Continuity and Transformation in Environmental Regulation Farber, Daniel A. 10 Ariz. J. Envtl. L. & Pol'y 1 (2019)

E.C. Fisher, P. Pascual, W.E. Wagner, Understanding environmental models in their legal and regulatory context. J. Environ. Law 22(2), 251–283 (2010)

S.I. Gass, M.C. Fu, *Encyclopedia of Operations Research and Management Science* (Springer, New York, 2013)

G. Gigerenzer and W. Gaissmaier, Heuristic Decision Making, Annu Rev Psychol. 2011;62: 451–82. https://doi.org/10.1146/annurev-psych-120709-145346

Guha-Sapir, D., Below, R. Hoyois, Ph., EM-DAT: International Disaster Database – www.emdat.be – Université Catholique de Louvain, Brussels, Belgium, (2009)

K. Hewitt, L. Sheehan, *A Pilot Survey of Global Natural Disasters the Past Twenty Years. Natural Hazards Research Working Paper, No. 11* (University of Toronto, Toronto, 1969)

E. Ippoliti, C. Cellucci, *Logica*, 2nd edn. (Egea, Milano, 2018). (in Italian)

J. Jaccard, J. Jacoby, *Theory Construction and Model Building* (Guilford Press, New York, 2010)

D. Kahneman and A. Tversky (1984) Choices, Values, and Frames. American Psychologist, 39, 341–350. https://doi.org/10.1037/0003-066X.39.4.341

T. Kuhn, *The Structure of Scientific Revolutions* (University Chicago Press, Chicago, 1962)

I. Lakatos, *The Methodology of Scientific Research Programmes* (Cambridge University Press, Cambridge, UK, 1978)

L. Magnani, W. Carnielli, C. Pizzi (eds.), *Model-Based Reasoning in Science and Technology, Abduction, Logic, and Computational Discovery* (Springer, Heidelberg, 2010)

I. McLeod, http://demonstrations.wolfram.com/AnscombeQuartet (2011)

MITRE Corp, *System Engineering Guide* (Washington, D.C., 2014)

Munich RE Foundation 2006, From Knowldge to Action, http://www.munichre-foundation.org/content/dam/munichre/foundation/publications/2007_1_en_Annual%20report%202006.pdf/_jcr_content/renditions/original./2007_1_en_Annual%20report%202006.pdf

NAS, Attribution of Extreme Weather Events in the Context of Climate Change, National Academic Press, Washington, D.C., (2016a)

NAS, How Modeling Can Inform Strategies to Improve Population Health (workshop Summary, National Academic Press, Washington, D.C. (2016b)

NAS, Characterizing Risk in Climate Change Assessments: Proceedings of a workshop, National Academic Press (2016c)

Office of Economic Cooperation and Development (OECD) Disaster Risk Assessment and Risk Financing, A G20/OECD Methodological framework, Paris (2015)

R. W. Perry, E. L. Quarantelli, *What Is a Disaster?*, 442 pp, (2005) (Rutledge, London Xlibris, 1998)

H. Rodriguez, E. L. Quarantelli, R. R. Dynes, *Handbook of Disaster Research* (Springer, 2007), 611 pp

Disaster Crisis Management: A Summary Of Research Findings, E. L. Quarantelli First published:July (1998) https://doi.org/10.1111/j.1467-6486.1988.tb00043.x

U.S. Environmental Protection Agency. (2015). *Climate Change in the United States: Benefits of Global Action*. EPA 430-R-15-001. Washington, DC. Available: https://www.epa.gov/sites/production/files/2015-06/documents/cirareport.pdf

UNDP, Human Development Report 2011, Sustainability and Equity, Paris France, https://www.undp.org/content/undp/en/home/librarypage/hdr/human_developmentreport2011.html (2015)

United Nations Development Program, (UNDP) Bureau for Crisis Prevention and Recovery, Disaster-conflict Interface, Comparative Experiences, NY (2015)

US Army Technical Report TR-2014-19, Army Independent Risk Assessment Guidebook, Aberdeen Proving Ground, MD, 21005, Washington, DC (April, 2014)

US EPA, Guidance for PM2.5 Permit Modeling, EPA-454/B-14-001, May 2014. https://www3.epa.gov/scram001/guidance/guide/Guidance_for_PM25_Permit_Modeling.pdf

WHO (World Health Organization). Definitions: emergencies. http://www.who.int/hac/about/definitions/en/

WHO/EHA, Disasters and Emergencies, Definitions, Training Package, Addis Ababa, (Updated March 2002). Addis Ababa, Ethiopia

Chapter 4
Catastrophic Incidents: Critical Information and Knowledge

4.1 Introduction

Public decision-makers should normatively make principled and rational choices because resources are scarce and have multiple competing claims over them. The spatial and temporal dimensions of prospective catastrophic incidents are critical to informing those choices. For example, the EU Commission (2014), after their review of natural and human-made disasters, found that specific time horizons are essential to define catastrophic events. The Commission stated that: "[a]s set out in the guidelines on risk assessment and mapping prepared by the Commission, national scenario-building and risk identification would need to consider at least all significant natural and man-made hazards that would occur on average once or more every 100 years and for which the consequences represent significant potential impacts, i.e. number of affected people greater than 50, economic and environmental costs above €100 million, and political/social impact considered significant or very serious."

The EU Commission adds (citations omitted) that: "[t]he timeframe for most of the risk assessments submitted is set at five to ten years ahead. This time period allows for a more reliable assessment of the probability of natural and man-made hazards ... and corresponds approximately to timescales for the funding... . ... a defined shorter timeframe may also help reduce comparability issues for risks which are important in the shorter term as compared to risks which may materialize only in the longer term."

We think that these time intervals are appropriate for assessing some, but not all, types of catastrophic incidents. CRED (Guha-Sapir et al. 2016, 2017) states that a catastrophic incident may be "a situation or event that overwhelms local capacity, necessitating a request at the national or international level for external assistance; an unforeseen and often sudden event that causes great damage, destruction and human suffering." This event is by no means similar to a routine event: it does not follow that a short time period of time increases the accuracy of the prediction

© Springer Nature Switzerland AG 2020
P. F. Ricci, *Analysis of Catastrophes and Their Public Health Consequences*,
https://doi.org/10.1007/978-3-030-48066-0_4

unless the mechanisms are well known. It is not surprising that there are different definitions and classifications of disasters. For instance, the US Centers for Disease Control (CDC 1989) identified three major categories of disasters: geographical events such as earthquakes and volcanic eruptions; weather-related events including hurricanes, tornadoes, heat waves, and floods; and human-generated ones. The latter of these includes famines, pollution from fossil fueled power plants, transportation, industrial disasters such as refinery fires, and nuclear power plant incidents such as civilian use of nuclear power to generate electric power. A number of human-caused accidents have been defined as technological (e.g., transportation, fires, explosions, chemical and radioactive release). The *source* of the consequences may range from *diffuse* (e.g., tens of coal-burning power plants in a region) to *centralized* (e.g., a very large hydroelectric large dam or several nuclear power plants in a single location). These two attributes are critical as is the interval of time within which casualties (and other consequence) occur. For example, a traffic accident at a specific locality and point in time may kill five persons. However, traffic accidents in the USA have recently, in 2014, killed approximately 32,000 persons; in the same year in the EU there were approximately 26,000 deaths. The temporal aspects of catastrophic incidents range from immediate or rapid (e.g., the overnight release of isocyanate) to slow (e.g., the continuous release of inorganic arsenic from natural sources to shallow aquifers from which people pump their water). Natural disasters cause damage against which there may be little direct recourse, for example, compensations, low interest loans, low cost insurance may not be available to those affected. Those who cause or otherwise illegally contribute to technological accidents can be prosecuted, subject to pay reparation for the harm done, fined, and perhaps jailed depending on the legal severity of the act: criminal or civil, volitional, or accidental. Latent disasters may go unrecognized for some time (e.g., large-scale exposure to chemicals not known to cause chronic effects when the manufacturing facility was built; exposures that seemingly cause clusters of increased prevalence of cancer). Disease clusters often defy scientific research that attempts to find a causal explanation because their cause may be confounded by other risk factors or be due to an hidden natural variablity. Although the number of adverse outcomes, namely the prevalence number (i.e., the total number of deaths or cases in a short period of time) associated with the disease, is a strong signal relative to the average (namely, the average background numbers of cases observed historically), the actual cause of the cluster may not be determined even after years of study.

The EU Commission (EU Commission Staff Working Document, Overview of natural and man-made disaster risks in the EU, SDW (2014) 134 Final (Brussels 2014)) considered events with known but low return periods: those are characterized as *significant natural and man-made hazards* and with relatively low probability of occurring. It included bounds on the magnitude of the consequences, accounted for qualitative concerns implying that the time periods are consistent with developing protective measures. Human-made hazards may be thought of as a set of either initiating events or as a cascading set of different events. For example, an earthquake causes its own direct adverse consequences but also one or more industrial accidents and loss of critical infrastructure that amplify the magnitude and characteristics of the overall consequences. There are many natural and man-made hazardous events. In Europe, for example, there can be volcanic eruptions, such as one that may happen in Italy and place hundreds of thousands at risk and near Naples. Chemical plants can, through human errors, release toxic gases that can poison or otherwise endanger thousands of individuals. Catastrophic incidents range from the *sudden non-repetitive* to *seasonal* (Banks 2005). We classify their spatiotemporal aspects in Table 4.1.

These aspects account for the evolution of disasters. Past information, the historical record, includes the magnitude of the consequences as a function of time and geographical location. Time accounts for the predicted or actual time of occurrence

Table 4.1 Four summary aspects of natural and human-made catastrophic events by spatiotemporal characteristics

Spatiotemporal locality aspect	Aspects of the catastrophic event	Examples and characteristics
Sudden	Occurs without discernible temporal and spatial patterns. Will generally not reoccur in the same location and yield the same results. Extremely large societal costs and loss of life.	Complete failure of large earth dam; technological disasters such nuclear power plant's loss of coolant accident (LOCA); multi-area fires. The gravity of the secondary impacts may not be fully anticipated: effects may be intergenerational and international.
Irregularly occurring	No known spatial and temporal regularity. Mechanisms may be known but their modeling is incomplete. Extremely large societal costs and loss of life.	Tsunamis generated by landslides: different locations can be affected differently. Mechanisms and warning signals may or may not be known. Effects may be intergenerational.
Mixture of seasons of various temporal lengths	Spatiotemporal predictable event according to some wave-like characteristics. Are measurable and known from historical data. Mechanisms are understood; the gradual accumulation of forces eventually triggers the event. Large societal costs and loss of life are known to happen.	Volcanic eruptions: re-occurrence may be measured in decades, centuries, millennia. Effects range from local to worldwide. Occurrence is determined by studying the geological record and from historical records. Technological disasters such as refinery fires. Generally, have much less than generational impacts. *Perfect storm* type of joint physical events creates unlikely extremes.
Seasonal	Occurs in a known location with known cycles. The *season* may be short (e.g., yearly) or longer (e.g., decadal). Mechanisms are well understood and so is their historical record. Large societal costs and number of casualties are possible, but the magnitude of the prompt fatalities is relatively low.	Hurricanes, floods, and droughts: occurrence is predictable from the recorded history and from observable physical phenomena. Events are consistent with the length and type of season (e.g., wet) or some multiples of it.

[a]*Season* means a regularity with predictable cycles; cycles may be year or decade long. We assume that a generation is 20 years

of the catastrophic incident. Its spatial domain, the source of the event itself, may be idealized as an area, line, or volume in some physical setting such as below ground, in shallow waters, in the atmosphere, and so on. The *location* of the event may be a complicated three-dimensional object, a volume (three physical dimensions: i.e., longitude (length), latitude (width), and depth or height (e.g., relative to mean sea level)). The main forces that these events generate, from earth movements to combustion, include pressure, shear, thermal radiation, and so on. The combined effects of the incident can be formalized as measures of well-being over time and distance from the *centroid* of the event. For example, the consequences of an event may be idealized to be like a cone centered at the location of the incident, the point of maximum damage. It may be attenuated by the physical distance from that point toward the edges of the cone: the *average* magnitude of the consequences decreases as a function of distance. Recurrent events cause consequences where both magnitude and frequency fluctuate over time: a cycle that may be thought of as a wave with constant period and amplitude.

The consequences are the salient elements of a catastrophe. An example, the WHO (who.int, accessed Dec. 4, 2019) estimates that the warming and precipitation trends over the past 30 years have caused over 150,000 deaths annually. The death tolls from blizzards are similar in magnitude to that of avalanches, roughly killing from 100 to 4000 per event (in the last 125 years or so); cyclones or hurricanes have killed up to 500,000 per event (in China, Bangladesh, and India) from 1876 to 2008. Famine has killed from more than 40,000,000 (China, 1958–1961) to less than 30,000 (Bangladesh, 1974, although reports indicate up to 1000,000 deaths). Famine can cause more than 1000,000 deaths per event; the actual magnitude may have been even greater than 10,000,000 deaths. Major earthquakes have killed 830,000 people (in 1556, Shanxi, China) and 10,000 (in San Juan, Argentina, 1944). Floods and landslides have caused approximately 3.7 million deaths (upper bound) in 1931 in China, and 100,000 in Vietnam, in 1971. Summer heat waves have killed 70,000 people in Europe (in 2003) and 739 (in the USA, in 1995).

Some natural catastrophes may be cyclic; others may have recurrence in periods measured from decades to millennia and more. An extreme example is super volcano eruptions, defined as magnitude 8 or greater VEI (volcano explosivity index), consisting of deposits from the eruptions that are greater than 1000 cubic kilometers (240 cubic miles). Those eruptions, in the past two million years, have occurred in what is now Yellowstone National Park, in Wyoming, Montana, and Idaho (approximately 9000 km²; its caldera is approximately 530 km²). Other super volcanoes are found in Long Valley, eastern California, Toba in Indonesia, Taupo in New Zealand, Japan, Italy, and South America. The Lake Toba super volcano eruption, about 74,000 years ago, is reported to have destroyed more than 99% of the human population (reducing it from 60,000,000 to less than 10,000), although these reports are disputed. The most recent super volcano eruption occurred 27,000 years ago at Taupo in New Zealand. According to the USGS (http://volcanoes.usgs.gov/vsc), the implications for the future are that: "Although it is possible, scientists are not convinced that there will ever be another catastrophic eruption at Yellowstone. Given Yellowstone's past history, the yearly probability of another caldera-forming eruption could be calculated as 1 in 730,000 or 0.00014%. This probability is roughly similar to that of a large (1 kilometer) asteroid hitting the Earth. Moreover, catastrophic geologic events are neither regular nor predictable ….." The USGS also suggests that "[m]ost scientists think that the buildup preceding a catastrophic eruption would be detectable for weeks and perhaps months to years."

In the last 100 years, volcanic eruptions have killed from more than 90,000 people (Mount Tambora, Indonesia, 1815) to more than 5000 (Mount Kelut, Indonesia, 1919). Almost two thousand years ago, in Italy, Mount Vesuvius eruption, in 79 AD, killed approximately 33,000 people (it has been reconstructed to have a VEI equal to approximately 6). Notably, if this volcano were to erupt again, it might kill up to 10,000 people depending on its VEI. Table 4.2 summarizes reports from national and international sources for earthquakes, floods, cyclones, avalanches, and communicable diseases, measured by the number of fatalities.

In 1995, the US Office of Technology Assessment (OTA, no longer in existence) reported that, by 2020 the probability of an earthquake of magnitude ≥ 7 (Richter scale) for the San Francisco Bay Area was 67%. For Los Angeles, the probability was between 80% and 90%, by 2024. In 2018, the USGS (usgs.gov, accessed Sept. 4, 2018) gives the following probabilities. For the San Francisco area, by 2030, there is a 31% probability of a magnitude 7.5 earthquake and a magnitude 6.7 earthquake has a 60% probability of occurring. For the same time period, the Los Angeles area has 31% probability to experience a 7.5 magnitude earthquake; the 6.7 magnitude has probability 60%.

The OECD (2012, their Fig. 2) summarizes the losses due to disasters in G20 and five non-OECD countries as percentage of averaged GDP over the 1980–2011 period, as a measure of economic loss from past catastrophic incidents. Chile was the country with the highest percent loss (~1.15%); the Russian Federation had the least percentage loss (~0.15%). The USA suffered approximately 0.20% loss placing it approximately in the middle of the 25 countries the OECD assessed. This agency (their Fig. 3) also depicts the distribution of the cost (in 2012 constant USD) of disasters, as a percentage of GDP, by World Bank income classes, for the same period of time. High-income countries suffered much lower (percent) losses than low-income countries, across all years, other than the period 1995–2000 when the averaged losses suffered by lower middle-income countries surpassed low-income countries. The key issue is that those who have the least resources suffer the most.

4.2   Spatiotemporal Characteristics of Catastrophic Incidents

For catastrophic incidents, we simplify our thinking by considering them as releases of forces that adversely affect us and our environment. An issue is the asymmetry between the temporal aspect of the feared future catastrophe and the almost current cost of prevention and protection. This asymmetry may create a *Faustian bargain*: policymakers balance a situation where thousands of individuals are at risk of death and injury by not spending sufficient moneys to protect them because the catastrophic

Table 4.2 Worldwide data on death burdens from earthquakes, floods, cyclones, avalanches and communicable diseases from various public sources

Earthquakes, floods, cyclones		Avalanches		Communicable diseases	
Deaths	Events, location, year	Deaths	Events, location, year	Deaths	Events, location, year
1,400,000 possibly 4000,000	Floods, China, 1931	20,000	From earthquake, Peru, 1970	300,000,000	Smallpox, worldwide, 1900 to eradication in 1980
900,000–2,000,000	Floods, China, 1887	265	Switzerland, Austria, 1951	200,000,000	Measles, worldwide, last 150 years
830,000	Earthquake, china, 1556	172	Afghanistan, 2010	100,000,000	Black death, Asia, Europe, Africa, 1300–1720
500,000	Cyclone, India, 1970	125	Russia, 2002	80,000,000 to 250,000,000	Malaria, worldwide, twentieth century to present
300,000	Cyclone, India, 1839	102	Pakistan, 2010	50,000,000 to 100,000,000	Spanish flu, worldwide, 1918–1920
230,000–310,000	Tsunami, Indian Ocean, 2004	96	USA, 1910	40,000,000 to 100,000,000	Plague in Asia, Europe, Africa, 540–590
250,000–300,000	Earthquake, Turkey, 526	90	Canada, 1910	40,000,000 to 100,000,000	Tuberculosis, worldwide, twentieth century to present
240,000–665,000	Earthquake, China, 1976	62	Canada, 1903	30,000,000	AIDS, worldwide, 1981 to present
234,117	Earthquake, China, 1920	59	Turkey, 1993	12,000,000 (?)	Bubonic plague, worldwide, 1850–1950
230,000	Earthquake, Syria	57	Austria, 1954	5000,000	Antonine plague, Rom, 165–180
142,000	Earthquake, Japan, 1923	–	–	4000,000	Asian flu, worldwide, 1956–1958
138,000+	Cyclone, Burma, 2008	–	–	> 250,000 per year	Seasonal flue, worldwide, (data from 2009)
138,000	Cyclone, Bangladesh, 1991	–	–	–	–
123,000	Earthquake, Italy, 1908	–	–	–	–
110,000	Earthquake, Turkmenistan, 1948	–	–	–	–

– Means no information

incident is predicted to occur far in the future. That money is spent on other beneficial actions unrelated to the hazard at issue.

Dilley et al.'s (2005) *Natural Disaster Hotspots: A Global Risk Analysis* rank disasters on an integer scale from the least, 0, to the worse, 10. These values are deciles of the distribution of location-specific consequences such as mortality and economic loss (gross domestic product, GDP). Using known and internationally defined indices, such as GDP, provides relatively robust indicators of the magnitude of the monetary consequences. The coping strategies that countries adopt blend local, national, and international mutual support. Societal stresses from catastrophic events can overwhelm the scarce resources a country has, deepen its poverty, increase dependence on foreign help, and result in short-run policies that can be counterproductive. Catastrophes involve damages and consequences to humans: in other words, a landslide that does not, directly or indirectly, affect humans may fall outside the direct concern of policy makers. This is not to say that those catastrophic events that may not affect humans, or are inconsequential to them, or are not relevant to an overall assessment. Even though humans may not be affected, other species may be at risk and those catastrophes are inherently relevant. Catastrophic events may occur as a sequence of events, each adding to the overall magnitude of the consequences. Prospectively, a chain of events may be triggered by an initiating event. In this case the probabilities are conditionalized on the preceding events (the chain can be described by a Bayesian network that we discuss in Chap. 5). If, on the other hand, the concern is with two joint events, and each event is characterized by a probability and the events are independent, their joint probability obtains from multiplying the individual probabilities. For example, if A has pr = 0.05 and B probability = 0.01, then their joint probability, pr(A AND B) = 0.0005. This is a different result from the sequential probabilistic analysis done with Bayesian networks because of the effect of conditional probabilities (i.e., the events in a network are dependent and thus require more complicated probabilistic analyses).

4.2.1   Magnitude, Location and Time

In 2004–2005, an earthquake-generated tsunami crashed against the shores of the Indonesian island of Sumatra, Sri Lanka, India, and Thailand; hurricane Katrina hit the US Gulf Coast and the city of New Orleans. The tsunami killed approximately 226,000 people; Katrina caused approximately 1850 deaths. The monetary costs from the tsunami were 8 billion USD; those of Katrina 125 billion USD. Both events generated large masses of water resulting in floods (and hurricane force winds, for Katrina). The deadliest catastrophes may not be the costliest. This is a stark reminder that mortality numbers and monetary costs are insufficient to characterize the absolute impact of catastrophes. These two very different catastrophic incidents bring into sharp contrast the nature of the forces that generate catastrophes far from their origin. The tsunami submerged entire coastlines of Indonesia within

Table 4.3 Dimensions are the units associated with each hazard

Hazard	Time	Dimensionality, units
Cyclones	1980–2000	Wind speed, kilometers per hour
Droughts	1980–2000	30% below standardized precipitation, 3-month average
Floods	1985–2003	Magnitude of flood events
Earthquakes	1976–2000	Frequency > 4.5, Richter scale
Volcanoes	1979–2000	Counts of volcanic activities
Landslides	NA	Counts of landslides and snow avalanches

minutes of the earthquake without warning. New Orleans was warned several days before Katrina affected it. Moreover, although hurricanes had affected New Orleans since it was settled by France in 1700 (and built below mean sea level) before the USA became a country, the evidence for Sumatra indicates that tsunamis had not occurred along its coast since 1500. Table 4.3 contains a summary of key characteristics of selected catastrophic outcomes caused by past natural hazards.

Against this background, we briefly discuss the unequal distribution of deaths by income group. The UNISDR, Centre for Research on the Epidemiology of Disasters, CRED, has developed estimates of the deaths by country from 1996 to 2015. It finds that more than 1.45 million people were killed by natural hazards during that period, most of which affected low- and middle-income countries. Of those deaths, 357,000 are attributable to earthquakes and tsunamis. In this period of time, there were three *mega-disasters*: the Indian Ocean tsunami (2005), Haiti's earthquake (2010), and cyclone Nargis (2008), each of which caused more than 100,000 deaths. In the period 1996 to 2015, 46.6% of the deaths occurred in low-income countries (627,232 deaths), 21.7% in the lower middle-income countries (292,789 deaths), upper middle-income countries suffered 22.4% of those deaths (301,469 deaths), and higher-income countries suffered 9.3% deaths (124,706 deaths) (UNISDR 2016). According to the Asian Development Bank (ADB), it appears that: "Asia and the Pacific have borne the brunt of this alarming trend: natural disasters are now four times more likely to affect people in the region than those in Africa, and 25 times more likely than those in Europe. A climate change vulnerability index indicates that all seven cities globally classified as being at 'extreme risk' are in Asia: Dhaka, Manila, Bangkok, Yangon, Jakarta, Ho Chi Minh City, and Kolkata. Financially, Asia accounts for almost half of the estimated global economic cost—close to $1 trillion—caused by natural disasters since the early 1990s."

Aspects of the economy (in US dollars) centered on year 2000, for several countries (Data from The World Bank 1999, their Table 1), are measured by the gross national product (GNP) and other economic attributes such as per capita GNP adjusted by purchasing power parity (PPP), yearly average growth rate of the GNP, population, and population density, Table 4.4. These data provide a relative understanding of the macroeconomic difference between India (~0.30% GDP losses), Turkey (~0.24% GDP losses) and the USA (0.21% GDP losses) discussed earlier. El Salvador and Mozambique are countries with much lower GDP and population.

Table 4.4 Summary of natural disasters data in 2011

Years	Average 2000–2009	2009	2010	2011	Location
Total number of disasters	392	335	385	302	Japan, Philippines, Brazil, Thailand, Turkey, Pakistan, USA, Cambodia, PRC, India
Total fatalities	78,087	10,656	297,000	29,782	Tohoku earthquake/tsunami in Japan, Philippines tropical storm Washi, and Brazil's mudslides. Floods in PRC, SE Asia, and Pakistan
Total economic damage, billions USD	89.3	41.3	123.9	366	See discussion in text

Source: EM-DAT: The Emergency Events Database – Université Catholique de Louvain (UCL) – CRED, D. Guha-Sapir – www.emdat.be, Brussels, Belgium. Source: EM-DAT, Natural Disasters, 2011

Economic data imply aspects of being able to cope with disasters and can be linked to economic losses associated with earthquakes, hurricanes, floods, and droughts in these countries. Although other factors should also be considered, the size of the GDP and the size of the monetary consequences from natural disasters give a sense of the overall extent of economic damage from specific hazards. For instance, if the 1998 GDP of El Salvador is 11.2 billion dollars and the 2001 earthquake that hit this country caused approximately 1.2 billion dollars of damage, it is clear that the country's economy has literally decimated – regardless of the differences in the GDP between 1999 and 2000.

As Dilley et al. (2005) note, international data from natural hazards are indicative of the magnitude of adverse consequences. These authors "[w]hile the data are inadequate for understanding the *absolute* levels of risk posed by any specific hazard or combination of hazards, they are adequate for identifying areas that are at relatively higher single- or multiple-hazard risk. In other words, we do not feel that the data are sufficiently reliable to estimate, for example, … mortality from flooding, earthquakes, and drought over a specified period." On the other hand, comparative relative analyses are possible provided that adjustments for purchasing power, inflations, and other socioeconomic factors are made to assure the homogeneity of the data used in possible comparisons.

Quantitative definitions of disaster influence the numbers stored in public or private databases. For example, the CRED database, EM-DAT, requires for inclusion meeting at least one of the following criteria:

1. 10 or more people reported killed
2. 100 or more people reported affected
3. Declaration of a state of emergency.
4. Call for international assistance.

For example, for the period 2000 to 2011, the EM-DAT database provides very useful worldwide aggregates. The value of the economic damage, according to the EM-DAT, is estimated to be $366*10^9$ USD. Another database, developed by Munich Re, estimates that value to be $380*10^9$ USD (Chap. 2, 2011 Natural Disasters Reviewed). The data in Table 4.4 include 2011, the year when most of the world-wide catastrophic incidents occurred, affecting approximately 210 million people. These data include famine and drought in the Horn of Africa (e.g., Somalia).

The average number of fatalities from *wet mass* movements, floods, and storms, according to Chap. 2, 2011 *Natural Disasters Reviewed*, from 2001 to 2010, were 1002, 5614, and 17,236, respectively. The number of these disasters, over that period, was 20, 175, and 104 and affected approximately 0.38, 106.3, and 39 million people, respectively. Their causes range from increasing average global temperatures to increasing length, frequency, and intensity of heat waves, rainfalls, and wind speeds. Other contributory causes may be demographic patterns of conurbation and increasing population density in coastal areas. 2011 was a very costly year: Munich Re estimated the economic losses associated with natural disasters to be $380*10^9$ USD, greater than any in the past. For example, the Tohoku earthquake and tsunami in Japan caused losses that exceed $210*10^9$ USD; floods and landslides in the Philippines caused damages of about $10*10^9$ USD; the New Zealand earthquake cost $7.3*10^9$ USD; storms and tornadoes in the US cost $7.3*10^9$ USD; and hurricane Irene in the Caribbean cost $7*10^9$ USD. Technological and other human-made activities also cause very large damages. Plantinga, Corubolo, and Clover (2014) surveyed major industrial accidents worldwide (from 2003 to 2012). Their findings indicate that the USA, China, India, Germany, and South Korea are ranked between the lowest numbers of deaths while Nigeria, Paraguay, Honduras, and Bangladesh between the highest.

Development yields tangible and directly rewarding and visible structures, products, jobs, and so on. Preventive or precautionary interventions may result in hard solutions, for instance, engineered defenses that can be costly while being undertaken. The Asian Development Bank (ADB) (2013) provides an example of a soft solution. It states that: "[u]nder a program led by the Bangladesh Red Crescent Society, the country built an exemplary early warning system based on a volunteer force of more than 30,000 men and women to disseminate cyclone warnings through village-level focal points and to carry out rescue operations."

As a result, while Cyclone Bhola in November 1970 killed over 300,000 people, another storm of similar magnitude, in 1997, took 188 lives. These soft solutions, particularly as part of a portfolio consisting of the combination of soft programs and hard solutions, can greatly decrease the extent and magnitude of the damage from natural and human-made hazards. Increasing demographic pressure and increasingly limited resources affect the existing infrastructure. The latter's vulnerability is due to aging, poor maintenance, and overcrowding. Moreover, the frequency of natural disasters is on the rise. For example, Munich Re has determined that the annual number of Category 5 storms tripled between 1980 and 2008. The Emergency Events Database (EM-DAT) of the Centre for Research on the Epidemiology of Disasters has been used to determine that occurrence of natural disasters has risen by about 600%, in the last 60 years. According to the ADB (2013): "[t]his likely reflects fundamental changes under way in global climate and surging populations in low-lying megacities, such as Bangkok, Manila, or Jakarta, ... Massive population growth in Asia has forced millions to move to more marginal lands and coastal areas, away from historically economically active areas along rivers and canals. Naturally, this has left people much more vulnerable to droughts or storm surges from typhoons." In Asia, the frequency and counts of deaths have been estimated by the Centre for Research on the Epidemiology of Disasters – Emergency Events Database (EM-DAT: The Emergency Events Database – Université Catholique de Louvain (UCL) – CRED, D. Guha-Sapir – www.emdat.be, Brussels, Belgium, 2016) that we describe in Table 4.5.

Guha-Sapir, Philippe, and Below (2016) report that, in 2015, the worldwide effect of major natural disasters, principally earthquakes, tsunamis, and extreme temperatures, resulted in "... 22,765 deaths, a number largely below the annual average for years 2005–2014 (76,416), and ... 110.3 million victims worldwide, also below the 2005–2014 annual average (199.2 million) ... with estimates placing economic damages at US$ 70.3 billion, natural disasters costs were, in 2015, significantly below their decennial average of US $ 159.7 billion."

In 2016, Guha-Sapir, Philippe, Wallemacq, and Below (2017) estimate that "... the number of people killed by disasters (8,733) was the second lowest since 2006, far below the 2006–2015 annual average of 69,827 deaths. However, this takes into account two years with more than 200,000 people reported killed, mainly attributable to mega catastrophes: the cyclone, Nargis, in Myanmar in 2008 (138,366 deaths) and the earthquake in Haiti in 2010 (222,570 deaths). Yet, even after the exclusion of these disasters, the number of deaths in 2015 remains far below a recomputed 2006–2015 annual average of 33,733 deaths."

EM-DAT, whose data is country-specific, distinguishes between two generic categories of disasters – natural and technological. CRED (Annex 1) contains 70 operational definitions of disasters, from *airburst* to *winter blizzard* and extends them to include prion disease, seiche, radio waves, epidemics, and so on. The information includes disaster by year (including start and end dates) and assigned identifier; country or countries of occurrence; and whether disasters are natural or technological. Disasters are classified as geophysical, hydrological, meteorological,

Table 4.5 Comparison of deaths counts from major world-wide industrial catastrophic incident

Disaster	Frequency			Death counts		
	1980–1989	*1990–1999*	*2000–2009*	*1980–1989*	*1990–1999*	*2000–2009*
Hydro-meteorological	502	781	1215	85,537	242,539	203,303
Geophysical	115	167	210	11,597	96,859	447,724
Total	617	948	1425	97,134	339,398	651,027

Source: EM-DAT: The Emergency Events Database – Université Catholique de Louvain (UCL) – CRED, D. Guha-Sapir – www.emdat.be, Brussels, Belgium 2016

climatological, biological, and extraterrestrial and by subtype, flood, and flash floods. The database includes lives lost and missing people, physical injuries, traumas, or illnesses that need immediate medical treatment. Additionally, it includes houses either destroyed or heavily damaged. Economic damage is measured in current (rather than deflated) USD; the numbers and their descriptors are obtained from official national data.

Increases in the magnitude of consequences may be due to demographic changes and preferences for locations in or near rivers, hills, mountain sides, and coastal areas. These areas have historically attracted more people who then are at increased risk from tsunamis, inundations, and landslides. Engineering measures to deal with natural or human-made disasters may also have been developed from anticipated increases in the number of people at risk. However, the predictions on which these measures were built may have underestimated the actual increases and their locations. For example, Guha-Sapir, Philippe, and Below (2016) report that the increase in the number of *reported* natural disasters in 2015 was mostly due to a higher number of *climatological* disasters: 45 compared with the 2005–2014 annual average of 32. The number of meteorological disasters (127) was 2% above its decadal average (125), while the number of hydrological disasters (175) and the number of geophysical disasters (29) were both 9% below their 2005–2014 annual average. In 2015, the number of people killed by disasters (22,765) was the lowest since 2005, way below the 2005–2014 annual average of 76,416 deaths which, however, takes into account 2 years with more than 200,000 people reported killed, each time mostly attributable to major catastrophes: the cyclone Nargis in Myanmar in 2008 (138,366 deaths) and the Haiti earthquake of 2010 (causing 225,570 deaths). But even after exclusion of these disasters, the number of deaths in 2015 remains below a recomputed 2005–2014 annual average of 40,022 deaths. At a more detailed level, it appears that in 2015 earthquakes and tsunamis killed most people (9526) followed by extreme temperatures that resulted in 7418 deaths, the second highest number since 2005 but far below the peak of 2010 (57,064). On the other hand, the deaths from floods (3449) and storms (1260) were the lowest since 2005, below the 2005–2014 annual averages (5933 and 17,769, respectively). The distribution of the consequences on those at risk before and after a catastrophic incident is inextricable from the analyses we discuss. For example, in China, the 2008 Sichuan earthquake killed 270,000 and displaced almost five million people. Hurricane Matthew caused approximately 176,000 people to be evacuated. Of the 2.1 million affected, 600,000 were children. The casualties were approximately 500 deaths and 350 injured. In the context of an ex ante analyses, we can never be sure of the exact numbers; however, analysis helps to locate emergency supplies, potential sheltered areas, and so on. This is routinely done. But it is the scale of the consequences that overwhelms society. Fundamentally, the effects on those affected go beyond preparations, however extensive. Changes in the health status of those affected will differ from the ex ante assessments.

Next we use six physical and economic examples to summarize the critical mechanisms mediating between cause and effect, a representation of the relationship between hazard and its consequences, and a basis of their economic analyses. Although these examples can be changed and made much more complex, they provide a plausible overview. The examples can be manipulated to generate other examples; they are applications of Wolfram Mathematica® demonstrations. In principle, they apply to routine, non-routine *Black-Swans* and *dragon-king* catastrophic incidents and exemplify formal components of the *physics* (as tangible mechanisms) mediating between cause and effect relevant to understanding the central concerns of science policy.

4.3 Analysis of Catastrophic Incidents: Six Building Blocks

The effect on society of prospective catastrophic incidents can be assessed at two levels: *general* and *particular*. The *general* level simplifies to: (i) the physical systems or processes that generate the forces that cause adverse consequences; (ii) the type (e.g., natural event) and its probability of happening; (iii) the distribution of the amounts and costs of the damages associated with each event. The simplest measures of the direct tangible cost to society are the magnitudes of the casualties and economic losses of capital and land. The complete set of adverse consequences can range from archeological to zoological; we limit the discussions to common, fundamental components. We combine both tangible and intangible valuations measured (however imperfectly) by a suitable monetary unit (i.e., *monetized* consequences stated in USD). The assumption that losses and many other consequences can be transformed into a single measure of value, money, can be a gross simplification. Nonetheless, it allows an easily understandable and simple relative ranking. To exemplify the scope of these analyses we discuss and describe six aspects. These are (i) tsunami waves, (ii) hurricanes annual per capita economic costs, (iii) probabilistic analysis of economic of production, (iv) probabilistic relationship between river water stage and their return periods including extremes, (v) an economic production function and the relationship between supply and demand for that production, and (vi) an economic extreme value, the value at risk, VaR. Our choices are personal; they should however give a general view of physical and economic subsystems that can combine with the discussions in previous chapters. The *particular* level is about detailed and specific causal mechanisms leading to adverse events and their consequences. Their modeling may take place through exposure-response functions (i.e., damage in which dose is related to the probability of cancer or other diseases) and through the assessment of the costs and benefits of optimal or preferable choices. The level of analysis consists of sub-processes, for example, ranging from genetic mutations causing irreversible changes to tissues to the bending moment associated with a force (or load) on a steel beam resulting in its failure. These specialized mechanisms are not discussed because they are far too numerous to be included.

4.3.1 Building Blocks

We have chosen these building blocks assuming an idealized catastrophic event with cascading effects. These are prototypical examples of: physical mechanisms (wave energy), probabilistic extreme values for initiating events, an economic analysis consisting of supply, demand, production and financial values at risk of loss.

Tsunami waves generation and characteristics The movement of tectonic plates relative to each other is gradual. If friction builds up and the plates do not locally move, then energy builds up until a break occurs causing an earthquake. If the event occurs below a body of water, the sudden movement of large earth masses may generate a tsunami (in Japanese, a *harbor wave*). Tsunamis can be generated by landslides or other earth movements such as an underwater volcanic eruption. As tsunami waves hit the shoreline, they can cause coastal flooding, erosion, land subsidence, and debris flows on land. Different land formations influence the height of the waves. The time to arrival of the waves varies, but damage can occur rapidly and unexpected ways. As an example (not shown), Trinh (2011), http://demonstrations. wolfram.com/TsunamiStrikingALandscape/, depicts the spatial and temporal profile of wave height. We use a 3-D modeling of a tsunami using Chang (2011), http:// demonstrations.wolfram.com/MathematicsOfTsunamis/, Fig. 4.1. The simulation of the 2004 tsunami depicts the influence on wave propagation from anomalies of the Indian Ocean floor. A $t = 0$, waves are rising above the underlying seamounts, while at the end of the demonstration, they are being propagated faster in the low-lying areas and they are higher above the seamount's crest. The demonstration uses modeling with partial differential equations (PDEs) for shallow waters associated with several physical variables, for example, describing the pressure surface of the water. A key assumption for this modeling is that the Indian Ocean is shallow: its depth is smaller than the open ocean wavelength of the tsunami.

Hurricanes Hurricanes are cyclonic systems with closed circulation around their centers that usually occur in the open oceans with warm waters. When a hurricane approaches a coastline, it causes very fast wind and storm surges that produce severe physical damage, economic losses, and casualties. Chandler (2011) shows hurricane risk by selected states of the USA. We use Texas (TX) and produce the direct per capita annual economic losses, in billion USD, from hurricane effects using his demonstration, http:/demonstrations.wolfram.com/HurricaneRiskByState. We depict the distribution of annual economic losses according to a two-parameters empirical Weibull distribution from data from 1900 to 2005, Fig. 4.2. The histogram plots log(probability) units versus magnitude; the actual Weibull distribution that can be fit to the histogram is not shown; the probabilities are historical frequencies.

Extreme value forecasting We use Bradley (2015), http://demonstrations.wol-fram.com/ExtremeValueForecasting/, for the Mississippi River station at Vicksburg to return periods and their 95% confidence bounds about the curve. The dataset

Event at t = 0

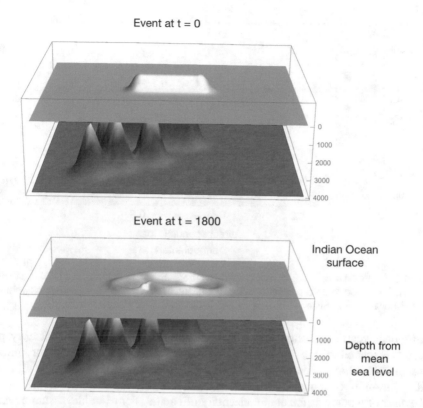

Event at t = 1800

Indian Ocean
surface

0
1000
2000
3000
4000

Depth from
mean
sea level

Fig. 4.1 The 3-D characteristics of the 2004 tsunami Pacific Ocean tsunami that killed over 200,000 individual is simulated using YS Chang (2011) http://demonstrations.wolfram.com/MathematicsOfTsunamis/ at $t = 0$ and $t = 1800$ units of time

depicts the maximum yearly height of Mississippi River from 1914 to 2014 using the *block maxima* approach applied to the extreme value theory, EVT, distribution, Fig. 4.3. The observed returns (black dots) are measured in years, and the gray and orange dots are predicted. The blue curve is the EVT cumulative distribution. The largest observed value is connected to its (maximum) return period by a dashed black line. The 95% confidence bounds are represented by two red lines about the blue line. They flare out to indicate that uncertainty increases the farther parametric curve is forecast away from the sample of data (observed over 70 years).

The next two Demonstrations describe the basic tools of economic analysis in terms of the change in economic output. Although a disaster affects many productive sectors of the economy, the effect in the change of the factors of production can be modeled by a production function. At the macro-economic level, the losses can be stated as changes in either total output, for example, GDP, or per capita output, GDP/person ($/person). We have used these elsewhere, as national yearly statistics. At the level of the single unit of production, the output can be modeled with the well-known Cobb and Douglas (short run) *production* function: $Y_t = A_t L_t^\beta K_t^\alpha$. In

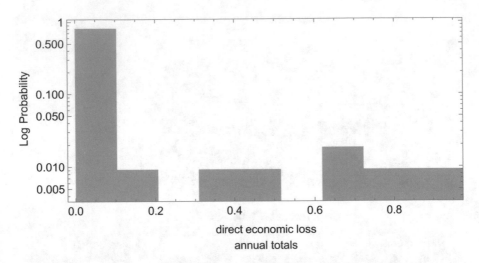

Fig. 4.2 Empirical distribution of the annual direct monetary losses in billions of USD due to hurricanes, from 1900 to 2005 computed using SJ Chandler (2011), http://demonstrations.wolfram.com/HurricaneRiskByState

this model, yearly economic output, Y_t, is measured in dollars or other currency, per year. It is related to (i) a dimensionless productivity factor A (accounts for innovation in an existing technology); (ii) labor, L, in units of persons-year; and (iii) capital, K, in units of money for capital equipment. The coefficient α is the output *elasticity* of labor; β is the output elasticity of capital. For example, assuming constant returns to scale, these two coefficients sum to unity: doubling the inputs doubles the output.[1] If the same production function can be used in both runs, we may be able to ex ante simulate changes to the economy due to a prospective disaster through changes in the factors of production.

Production function Following Kellenberg and Mobarak (2011), the dynamic form of this production function can be changed to a differential equation to obtain the trajectories of the changes in the output over time. The Mathematica demonstration of the static (not time-dependent) Cobb-Douglas production function that we

---

[1] Elasticity is defined (for $y = f(x)$) as the % change in the output, y, from a % change in the input x; using derivatives: $d(\log(f(x)))/d(\log x)$, $x > 0$ and $f(x)$ differentiable. Partial derivatives should be used for the Cobb-Douglass or any other continuous and differentiable production function with more than a single input. Clearly, as is the case for slopes that can change from point to point, so do elasticities (of output, price, income, and so on). Elasticity, as price elasticity of demand, is depicted by F. Maclachian, (2011) Revenue and Elasticity, http://demonstration.wolfram.com/RevenueandElasticity. No not worry that the units of the factors of production do not cancel out, relative to the output of the production function. The idea embodied in this function is what matters for it being a building block. Importantly, the production function for an assessment of the change in output should be associated with the long-run, rather than the much simpler short-run, the effect of time should be added, all physical factor of production became time dependent, and the technological coefficient would not be a constant.

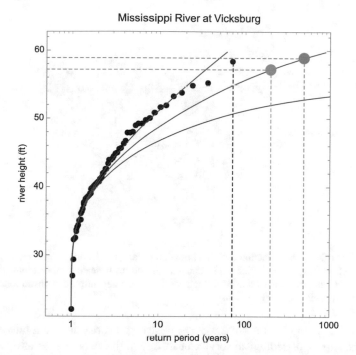

Fig. 4.3 Plot of return periods and statistical uncertainty estimated for the maximum yearly height of Mississippi River, from 1914 to 2014, using the block maxima method developed using M. Bradley (2015) http://demonstrations.wolfram.com/ExtremeValueForecasting/

depict in Fig. 4.4 was developed using SJ Chandler, http://demonstrations.wolfram.com/CobbDouglasProductionFunctions/.

Figure 4.4 is an example of the results that obtain from combining two inputs (l, labor, and k, capital, both different from 0 and a technological factor, $a = 1.50$, which scales the output, relative to $a = 1$ (i.e., doubling the inputs doubles the output)). A natural extension of the two-input function would include land (e.g., in meter^2). The theory is that the key factors of production are capital, labor, and land and that output is generated by them using a multiplicative rather than an additive function.

From this economic perspective, an output implies a supply and a demand for it, or else it would not be produced. The static, combined representation of demand and supply functions for goods and services give an intuitive understanding of how to analyze some salient economic effects of a catastrophic incident, at a specific point in time. This also implies that either supply, demand or both may disappear. Although there are many other effects, from public health to social, ecological, cultural, and many others, their analysis can generally be reduced to simple supply and demand analysis that may account for taxes, subsidies, and so on either over time or statically. This analysis can approximate the value of ecological services for which there is no market pricing mechanism as there is for normal goods and services we

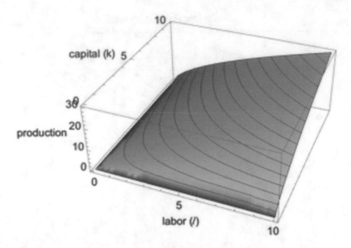

Fig. 4.4 Cobb-Douglas production function in which output, production, is generated by the production function $1.5*k^{0.65}l^{0.65}$, which implies increasing returns to scale for the output. The sum of the exponents determines the scale of the returns (SJ Chandler, http://demonstrations.wolfram.com/CobbDouglasProductionFunctions/)

use daily. For example, if a flood were to destroy an ecosystem, but have no other effect, the supply of ecological services lost due to the flood can be assessed using a production function. This approximation provides quantitative information that can be incorporated in the cost and benefit analysis of protective choices. Supply and demand curves for ecological services can be understood as follows. The demand curve can be developed by analogy to the demand of similar goods and services found in the economy. To the extent that there is an aggregate demand for goods and services that can be lost due to a catastrophic incident, the relationship between supply and demand, depicted in Fig. 4.3, provides the (static) price-quantity equilibrium levels needed to inform policy choices. This assessment comprises the cost side of the analysis: the supply curve is the producer's ability to supply of quantities of the good or service under analysis. Each supply and demand system applies to a single good or service, and its variables are price [$/unit demanded] and quantity [number of units produced].

Supply and demand The basic relationship between aggregate supply and demand is modeled using M Gillis (2011), http://demonstrations.wolfram.com/BasicSupplyAndDemand/. The orange line depicts the supply curve; the blue line depicts the demand curve, Fig. 4.5.

The two values, P^e and Q^e (the superscript e is not an exponent, it is an indicator for equilibrium value) correspond to the *equilibrium* point between price and quantity in a static economy, in a specific time period. As an example, in ecological resources management, this joint relationship can help understand the effect of a sudden disturbance on the supply side (e.g., destruction of a habitat). In this case, the supply curve disappears, and demand should in principle remain constant unless there were changes in the preferences by consumers or better alternatives became

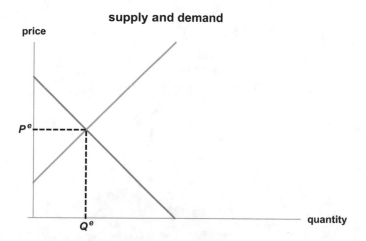

Fig. 4.5 Static and deterministic examples of economic models of aggregate supply and demand for normal economic goods and services, Gillis (2011) http://demonstrations.wolfram.com/BasicSupplyAndDemand/

available to them. In these conditions, the model is more complicated. A natural complication to understand the dynamics of the equilibria suggests looking at several static situations (over several time periods) before the disaster. The trace of the equilibria mapped by yearly couples of P^e and Q^e may approximate the dynamics of the economic system over time.

Value at risk (VAR)   Another important aspect of the effect of catastrophes, as economic losses, can be analyzed using the value at risk (VaR), Fig. 4.6, of an investment associated with precautionary choices. Although the context is narrowed to a direct economic loss, its principles are relevant when considering the allocation of scarce resource and the probability of some catastrophic loss of revenue. The uncertainty regarding the loss increases as time increases: the larger the VaR, the less certainty the outcome. The confidence level is indicated how the VaR drops within this confidence interval (90%). Nagy (2011) *Value at Risk* http://demonstrations.wolfram.com/ValueAtRisk/ depicts the characteristics of the VaR. In Fig. 4.6, value at risk, VaR, quantifies the probability of loss of an expected future sum. The term profit means a positive return on the investment. These terms are most likely to apply in private decision where profits are clearly defined, unlike public expenditures. The VaR quantity is a useful concept when alternative allocations may be considered.

4.3.2   Combining Analyses for Prevention: A System's View

The key elements of a policy-based discussion of catastrophic incidents suggest: (i) relating the magnitudes of the forces generated by an event to their probability of occurring; (ii) the damage function in which the magnitude of each force is related

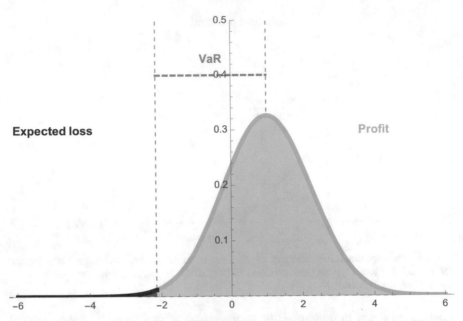

Fig. 4.6 Value at Risk, VaR, quantifies the probability of loss of a future investment, for example setting aside some money to be available in the future, for a given number of arbitrary time periods (Nagy 2011, Value at Risk http://demonstrations.wolfram.com/ValueAtRisk/)

to direct damage; and (iii) assessing the different opinions that scientists may have and how to aggregate them into a coherent whole that informs decision makers about magnitude of the costs of damage and those of prevention. Figure 4.7 depicts several aspects of these analyses and is an aggregate of established relationships found in the hydraulic engineering and hydrological literature. The relationship between the probability and magnitude of the force leads to the distribution of the damages, as local and direct consequences of the forces causing damage. The magnitude of the damage results in ex ante and prospective costs of prevention and costs of the damage. These costs are random variables and thus are described by a distribution function. The figure includes the dynamics of the evolution of the relationship between degradation to withstand the force generated by an event as well as the probable forces that may affect a structure at risk. The latter of these exemplifies that an existing structure is subjected to different forces over time. Those forces may increase relative to those assumed in the design-construction phase after a period of the economic life of the structure: the intersection between lines S and R, after time $T*$.

Figure 4.7 depicts what we consider to be a minimal set of relationships that characterize the critical elements linking cause and effect for prospective hazards. Although we use continuous lines, the information often consists of discrete data and thus requires statistical modeling to generate these lines. We make several simplifying assumptions. For example, although a structure is generally made of many components that distribute the loads to which the structure is subject, we lump these

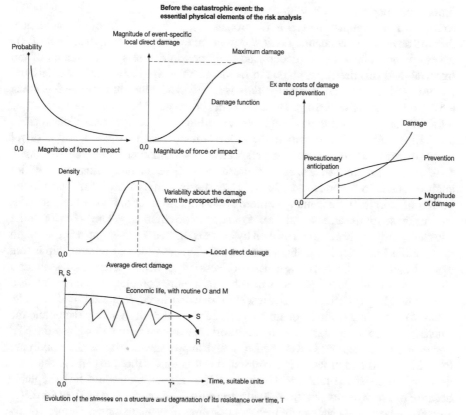

Fig. 4.7 Integrated relationships for the analysis of risks (e.g., probability and magnitude) of forces acting on structures, as a function of time; S stress, a force, R resistance, the resisting force, O and M are operation and maintenance). Economic life is assumed to be calculated from data, changes over time are hypothetical but related to ageing with normal O and M

into one: the structure. We assume that human losses can be monetized, as can all other forms of damage, and that proper discounting is applied to the value of money. There remains a key aspect of these analyses: How practically to deal with uncertainty when considering probabilistic events and their consequences. Engineering failures are of different types (e.g., buckling, yielding): the result is that the structure loses integrity. Failure can be due to poor or non-compliant design and construction. Over the life of the structure, poor maintenance, dynamic stresses unaccounted for, environmental factors such as corrosion, and so on affect its reliability. Its safety factors (SF) can be increased, but the increase causes higher costs and other issues, including aesthetic. The *limit state* approach implies that the *state* (either desirable or undesirable) of a structure is the point where a structure exceeds its operational capability. This state has a characteristic value; it is a portion of the distribution of failures associated with a specific stress. For example, given a structural element, a limit exists beyond which, with a small probability, fatigue adversely affects the structure: For example, there is a 5% probability, with a 75% level of confidence, that the adverse effect occurs. *Limit states* are intended to include the effect of: (i) excess (unaccounted for, but understood to be possible) stresses, (ii) quality of the material, and (iii) deterioration due to factors such as age of the structure, fatigue, erosion, and so on. In this context, *failure* is a statistical concept: for example, it has a 5% probability of occurring. These concepts correspond to the broader aspects of reliability analysis in structural engineering and in public health. Specifically, these are failure and survival distributions: failure means lack of success and survival means success. From a sampling perspective, experimental loads and resistance tests yield these frequencies. A very useful discussion of probabilistic reliability analysis is due to Bommel Ltd. (2001). They discuss safety factors, *SF*, used in the design of structural elements and, more generally, in other analyses including *SF* such as those for the assessment of exposure to toxic chemicals. The behavior of the structure being analyzed is established by past experiments and engineering strength of materials analyses. Regarding its structural elements, the key relationship is that a (static and deterministic) *design* load (or stress) is applied to determine a structural element's *design* characteristics. A member of a more complex structure such as a dam or building is designed to resist with a resisting force, R, a total load (a force, stress, S) according to the inequality $S_c \leq R_c/SF$, in which SF is safety factor and the subscript c implies a specific load condition that numerically determines the SF.[2] For example, a member is designed to resist approximately twice the maximum load, SF = 2. Although safety is increased, the structure is more expensive than one designed to resist the load for which the SF = 1.0. This may not be acceptable because of flaws, faulty practices, load variability, and so on. The units of force (or stress) can be pounds per square inch or kilograms per square meter. Failure behaviors depend on the materials used. For example, *elastic* behavior may be followed by *plastic* (deformation) and then by partial (e.g., spalling) or total failure. If the safety factor is 2, the structure is designed to support twice its design load as a function of elastic behavior: plastic behavior is not accounted, and thus the structure may be able to withstand more load before failure. A technical assumption is that a homogenous material, for example, steel, is elastic (can rebound to its initial state), rather than plastic (it deforms before failing). A load S_c can be underestimated or overestimated: S and R can be analyzed deterministically when there has been much experience with loads, materials used, quality assurance and control, QC, O&M, building practices, permit reviews, and so on. Loads are added, by type, and the SF calculated accordingly: $\sum_i^n \gamma L_i = S_d \leq R_d = R_k / \gamma$, in which γ is the load factor (< 1 or > 1), L_i is the type of load (e.g., live, dead). Failure itself may not be the sole criterion for safety and precautionary concerns: the unimpeded use of the structure during its economic life may be relevant. These issues are exemplified in Bommel (2001). These concepts suggest when simplifications from probabilistic to deter-

---

[2] The SF should account for requirements of building or other codes, manufacturing and construction practices in the field, and so on.

ministic approximations are appropriate. For example, if forces or energy that may be directed against a built structure that can be assumed to be deterministic, perhaps because of their past history, then treating those forces or energies (e.g., potential and kinetic energy) as deterministic may be codified. The simplification from a probabilistic view to a deterministic one may be based on the expectation that technological processes are known historically to improve. Thus, the simpler approach may be well-justified and can be improved by future changes in practices and in the technical information contained in building and other codes of practice.

4.4   Conclusion

The effect of catastrophic incidents tragically extends far beyond their human casualties. Although those numbers may be the most salient, the totality of the consequences is often unimaginable. From displacing millions of people to the loss of animals, houses, and land, from the loss of potable water and electricity to loss of employment, and from destroyed hospitals to the effects on the remaining infrastructure to support extreme demand for food and shelter, catastrophic consequences cascade in ways that may not be foreseen. In some cases, the loss of human lives may be relatively small or even zero; yet, the broader social and economic impacts can be enormous. At the national level, depending on the development of the country affected, the ability to rebound can be reduced and thus affect the country's future development and increase poverty. It is not a simple matter of numbers. For those affected, and for those who should protect them, there are duties that go beyond numbers and calculations. Yet, the duty should be informed by numbers and calculations because public funds are scarce and should be directed toward the maximum net benefit for society. An informed citizenry can more efficiently deal with disasters before they strike. The right to be protected implies the duty to be informed regarding the nature of what the citizens seek protection against.

These comments imply a concern with the equitable distribution of the consequences and the probability of being affected. This issue is exemplified by the differential impact of natural disasters on women and men. The critical finding is that women are more affected than men by reducing their life expectancy (Neumayer and Plumper 2007). This inequity correlates with socioeconomic status where poorer women suffer larger effects. As these authors report, the higher the socioeconomic status, the more equal the impact: men and women are almost equally affected. For women, the risk of death can increase even after the disaster due to prevailing social customs. For instance, in the aftermath of the 1995 Kobe earthquake in Japan, women were found to be at 1.5 higher risk than men.

References

Asian Development Bank, ADB, The Rise of Natural Disasters in Asia and the Pacific: Learning from ADI's Experiences, Mandaluyong City, The Philippines (2013)

E. Banks, *Catastrophic Risk: Analysis and Management* (Wiley, Hoboken, 2005)

Bomel, Ltd, Probabilistic Methods: Uses and abuses in structural integrity, UK HSE, CR Report 398/2001, 2001

M. Bradley, http://demonstrations.wolfram.com/ExtremeValueForecasting/ (2015)

S.J. Chandler, http://demonstrations.wolfram.com/HurricaneRiskByState (2011)

S.J. Chandler, http://demonstrations.wolfram.com/CobbDouglasProductionFunctions/ (2015)

Y.S. Chang, http://demonstrations.wolfram.com/MathematicsOfTsunamis/ (2011)

M. Dilley, R.S. Chen, U. Deichmann, A.L. Lerner-Lem, M. Arnold, et al., *Natural Disaster Hotspots: A Global Risk Analysis* (The World Bank, Washington, D.C., 2005)

EU Commission Staff Working Document, Overview of natural and man-made disaster risks in the EU, SDW (2014) 134 Final (Brussels, 2014)

M. Gillis, http://demonstrations.wolfram.com/BasicSupplyAndDemand/ (2011)

D. Guha-Sapir, H. Philippe, R. Below, *Annual Disaster Statistical Review 2015, The Numbers and Trends* (Centre for Research on the Epidemiology of Disasters (CRED) Institute of Health and Society (IRSS) Université Catholique de Louvain – Ciaco Imprimerie, Louvain-la-Neuve, 2016)

D. Guha-Sapir, H. Philippe, P. Wallemacq, R. Below, *Annual Disaster Statistical Review 2016, The Numbers and Trends* (Centre for Research on the Epidemiology of Disasters (CRED) Institute of Health and Society (IRSS) Université Catholique de Louvain – Ciaco Imprimerie, Louvain-la-Neuve, 2017)

D. Kellenberg, A.M. Mobarak, The Economics of Natural Disasters. Ann. Rev. Resour. Econ. **3**, 297–312 (2011)

F. Maclachian, Revenue and Elasticity, http://demonstration.wolfram.com/RevenueandElasticity (2011)

G. Nagy, http://demonstrations.wolfram.com/ValueAtRisk/ (2011)

E. Neumayer, T. Plumper, The gendered nature of natural disasters: the impact of catastrophic events on the gender gap in life expectancy, 1981–2002. Ann. Assoc. Am. Geogr. **97**(3), 551–566 (2007)

OECD, *Disaster Risk Financing, A global survey of practices and challenges* (OECD, Paris, 2012)

A.A. Plantinga, D.J. Corubolo, and R. Glover, Catastrophe modelling: deriving the 1-in-200 year mortality shock for a South African insurer's capital requirements under solvency assessment and management (SAM) Presented at the Actuarial Society of South Africa's 2014 Convention 22–23 October 2014, Cape Town International Convention Centre

Trinh,   http://demonstrations.wolfram.com/TsunamiStrikingALandscape/,   Wolfram Demonstrations Project (2011)

UNISDR, Center for Research on the Epidemiology of Disasters, CRED, Poverty and Death disaster Mortality 1996–2015 www.unisdr.org (2016)

WHO, Health and Environment Linkages Initiative (HELI), (who.int, Accessed 4 Dec 2019)

World Bank, *Development Report 1990/2000, Entering the 21st Century* (Oxford University Press, Washington, D.C., 1999)

Chapter 5
Uncertainty: Probabilistic and Statistical Aspects

5.1 Introduction

This chapter provides a simple probabilistic overview of the analysis of common, rare, and extremely rare events. Prospective catastrophic incidents can be viewed as changes from a normal state (e.g., a physical or economic system is at equilibrium with its environment) to a new terminal state, a partial or total failure. The change of state may be due to changes internal to the systems' mechanisms or to external (exogenous) inputs. More specifically, within a physical system that characterizes the path from an initiating event to its consequences, probabilistic uncertainty can be aleatory and epistemic. The former uncertainty is associated with, we colloquially call, *chance*. It prospectively describes one or more events that are not under the control of the experimenters or observers. Regardless of those individuals' efforts, an aleatory event is inherently unpredictable. On the other hand, the latter uncertainty is under the control of those individuals but requires effort in collecting and analyzing data; they entail costly and time-consuming efforts. Regarding probabilistic analyses, the NAS (2000) recommendation (to the US Army Corps of Engineers (USACoE), emphasis added) "to use *annual exceedance probability* as the performance measure in engineering risk" is exemplary for our discussions. Although it is stated in the context of flood-related natural events, the concept can be applied to other events in which probabilities (or frequencies) involve threshold values. For example, the USACoE considers the entire length of a levee and two flood-stage heights to calculate the *probable failure point* (PFP) (set at 0.85) (USACoE 1992, in NAS 2000, p. 68). This failure point is associated with an annual exceedance event. The assessment of the event's consequences, Public Law 104-303, October 12, 1996, Section 202 h, is such that that the USACoE should use "... risk-based analysis for the evaluation of hydrology, hydraulics, and economics in flood damage reduction studies. The study shall include an evaluation of the impact of risk-based analysis on project formulation, project economic justification, and minimum engineering and safety standards; and a review of studies conducted using

© Springer Nature Switzerland AG 2020
P. F. Ricci, *Analysis of Catastrophes and Their Public Health Consequences*,
https://doi.org/10.1007/978-3-030-48066-0_5

risk-based analysis to determine—(i) the scientific validity of applying risk-based analysis in these studies; and (ii) the impact of using risk-based analysis as it relates to current policy and procedures of the Corps of Engineers."

This citation is generalizable to most studies of natural and technological catastrophic incidents through changing terms (i.e., *hydrology*, *hydraulics*, *flood*) with context-specific alternatives. Of course, the USACoE may not be the appropriate agency. We discuss the critical components of this generalization: initiating events, mechanisms, and consequences. Uncertainty, described by probability measures, affects these elements of the analysis. It is combined through formulae, systems, and algorithms. The consequences depend on what is at risk of failure. For instance, a reinforced concrete column may fail if it is subjected to greater loads than the design load, as can an earth dam or more complex structure. The phrases *probability of failures* and, as an alternative, the *probability of survival* apply to either a single element or an entire structure. At one extreme, consider a landslide caused by a natural event: an uphill earth mass slides or otherwise separates from a mountain side perhaps due to excess rainfall. Once the stable state of that mass changes to unstable state the result may be catastrophic. At the other extreme, a population of normal cells, with some probability, may change state and become aberrant in such a way as to cause cancer. This change may be due to environmental exposures to a natural agent (e.g., radon gas in the air) or from a chemical in the workplace (e.g., arsenic in the water being drunk by a population). The transition probability from normal to abnormal can be modeled using probabilistic methods.

As we will discuss in more details, *information* implies *observations*, *data*, and *results* that are used interchangeably to mean primary quantitative information. A *result* obtained through a formula or model can be taken as raw information: that is, the output becomes the input into another formula. Inputs and outputs can be of various forms (e.g., discrete or continuous; deterministic or probabilistic). The term *knowledge* implies the use of these terms within a descriptive or predictive model. It is a system that represents the reality of interest; that reality can be *physical*, *conjectural*, or others. A *system* may imply simultaneous relationships (such as supply and demand), feedbacks, temporal lags, and so on. *Data* as inputs in statistical analysis are not further divisible. Data can be thought of forming three distinct subsets: one consists of the vast majority of the data (say 90% of them), while the two other sets contain the minima and maxima (in this example, a total of 10%). The percentage contained in the two sets of minima and maxima may be symmetric (e.g., 5% and 5%) or asymmetric (e.g., 1% and 9%). In the context of developing models, we will often be dealing with propositions connected by logical operations. For example, input *premises* are connected to *rules* by operators, such as *AND* or other logical operators, to yield the output as propositions that become *conclusions*. Uncertainty, ranging from imprecision to vagueness, pervades all calculations and the building of models. What matters is the nature of uncertainty and how it is described. For example, uncertainty may be thought to fall into two distinct categories which we interpret as follows:

- *Natural variability* – also termed *aleatory* – is an inherent state-of-nature of an object. This uncertainty cannot be reduced. However, knowledge changes as science advances: theoretical and empirical understanding of physical processes change. Natural variability may become better understood, because some particular aspect underlying it, at the end of a sufficiently long period of time, becomes better known.
- *Knowledge uncertainty* – also termed *epistemic* – is associated with the inability fully to formalize a process that generates data because of limited K&I. Practically, the upper and lower confidence bounds about an event's frequency-magnitude curve describe this uncertainty. This uncertainty is due to sampling, assumptions about the mechanism that generates the observations or data, and other technical issues. For instance, knowledge uncertainty may be about computable errors such as those made by fitting a model to data, choosing from alternative models, and forecasting future output values. Upper and lower bounds represent the probabilistic *confidence* selected by the analyst, *1-α*; where *α* is probabilistic *significance*. Data analyses determine the probability with which statistical sample estimates of a population parameter will converge, under repeated random sampling, about 1-*α* percent of the time, on the true population parameter value or values.

Physical and other phenomena can generate adverse consequences. Their representation uses a formal language, such as algebra, to express an analyst's understanding of the reality she seeks to represent and predict. A mechanism (or system if necessary) relates one or more inputs to one or more outputs (e.g., independent variables to dependent variables via mathematical operators). For example, in the deterministic function: *output = k*(input)*, *k* is a known number (a coefficient). Both input and output data are measured on the real axis, and if the coefficient *k* = 2.5, it is certain that *Y = 2.5*X*. Given this function, prediction and description are both simple and clear. Importantly, we have neither placed a limit on their values nor stated the units of their dimensions. Moreover, *X* causes *Y*, and vice versa. We can also think of a simple constraint on this model: neither *X* nor *Y* can be negative. More complicated constraints should easily come to mind, given a practical context and its model. If the relationship between these two variables were uncertain, the entire analysis would have to be changed to being statistical, as discussed throughout this chapter.

Example Exponential decay is often the basis of reasoning about effects that change over time. We use Wolfram's *Exponential Decay* demonstrations http://demonstrations.wolfram.com/ExponentialDecay/ to represent the deterministic reduction in the quantity of an object over time. We model the exponential decrease in a quantity *N*, for instance, a population [number of individuals], according to the expression (a differential equation):

$$\frac{dN}{dt} = -\lambda N \tag{5.1}$$

This model has solution: $N(t) = N(t = 0) * Exp(-\lambda t)$: this is the trajectory of the population changing over time, with a known initial population number N at time $t = 0$. The term Exp(.) equals $e^{(\cdot)}$, the exponential term, e is Euler's number, an (irrational) number that approximately equals 2.71828.... It is the base of natural logarithms (ln). Assuming a decay constant $(-\lambda = 0.5)$ and using S. Wolfram http://demonstration.wolfram.com/ExponentialDecay/ we obtain the plot in Fig. 5.1.

This formalism can be extended to more complicated issues associated with catastrophic incidents. The prediction of the spread of infectious diseases caused by biological agents, perhaps following a major natural catastrophe, is an important issue, particularly for events that generate cascading consequences. For example, consider modeling the spread of an infection in a population at risk. In this model, we use incidence (new cases per year) rather than prevalence (all cases per year) data. Suppose that the incidence rate of the disease in a period of time is $I(t)$, we want to calculate the time that it takes for the incidence to double, t_2. Let $I(t = 0)$ be the initial incidence and assume that the process increases at an exponential rate r [new cases/year]. We can describe this process as: $I(t) = I(t = 0)exp(rt)$. By substitution: $2*I(t = 0) = I(t = 0)exp(rt_2)$. Taking the natural logs of both sides of the expression and simplifying, we obtain: $t_{1/2} = ln(2)/r = 0.693/r$. Let $r = 0.05$ [cases/year]. Therefore, the doubling time is $0.693/0.05 = 13.86$ [years]. Now, let $I(t)$ be the *incident* cases in a population of size N. Let k [incident cases/year] be the rate at which the number of cases changes, assume that this rate is independent of individual behavior or other factors; and assume that *transmission* is direct. Let k be the relative growth rate: $[dI(t)/dt]/I(t)$. In a period of time dt, $I(t)$ will cause $kI(t)$ new

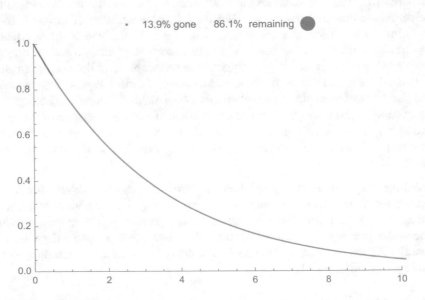

Fig. 5.1 Exponential decay of an arbitrary quantity. The x-axis is units of time and the y-axis the quantity N. N is decreasing at the constant rate 0.5 (S Wolfram http://demonstration.wolfram.com/ExponentialDecay/)

cases to occur. The fraction $I(t)/N$ of $kI(t)$ is already infected. The new cases in the period dt are $\{kI(t)-kI(t)[I(t)/N]\}$. It follows that $d[I(t)]/dt = kI(t)-kI(t)[I(t)/N]$. In the initial phase of the diffusion of the disease, $1-(1/N) \approx 1.00$: N is much larger than $I(t)$. Therefore, $d[I(t)]/dt = kI(t)$. This differential equation has the solution: $I(t) = I(t = 0)exp(kt)$. If we let $k = 2$ cases/year and the initial incidence be *five* cases, we obtain: $I(t = 1.5) = 5*exp(2*1.5) = 100.4$ [cases], at the end of *1.5* years. This modeling implies continuous compounding. If k is not constant but varies inversely with time, $k = m/t$, the model becomes: $dI(t)/dt = m[I(t)/t]$. The solution of this differential equation is $I(t) = I(t = 0)t^m$. The doubling time equals $\sqrt[m]{(2-1)}t$; the value of m depends on the disease and is determined empirically.

5.1.1   *Comment*

How can epistemic uncertainty about the value of k, λ and that of the coefficients of the model of the spread of an infectious disease be determined? The certain function (Eq. 5.1) now has to change to being probabilistic. For example, a linear correlation may be computed between input (i.e., x) and the output (i.e., y) to obtain an estimate of k. In the deterministic instance k was known with certainty. The transition from deterministic to probabilistic reasoning considers: (i) the form of the mechanism is assumed or hypothesized to be linear; (ii) the numerical value of k is obtained from data and the estimation of its value is based on several statistical assumptions and calculations based on sampled (from a population) *input* and *output* data on x and y, $x \epsilon X$, $y \epsilon Y$. The numerical value of k is no longer certain. Rather, it is contained in a probabilistic interval. The interval measures the probability $1-\alpha$ (this probability is the level of statistical *confidence*); α is probabilistic *significance*. Their sum equals *1.00*. For example, the 95% (0.95) statistical confidence level and the *0.05* statistical significance level. Because we are dealing with statistical analyses, we want to control the probability of accepting the null hypothesis when it is true. This probability is set on the basis of the implications of the research: it is data driven. For example, $pr(H = 0) = \alpha = 0.05$ means that the researcher wants to be sure that the error consisting of rejecting the true null (e.g., of no effect) hypothesis to be 5 %. This threshold probability is *1/20*. An alternative and lower value may be more appropriate depending on the context. These issues are formalized by the next example.

Example E Schultz demonstration http://demonstrations.wolfram.com/Decisions BasedOnPValuesAndSignificanceLevels/ produces the results that follow: (i) *reject H₀ in favor of Hₐ*; immediately below, (ii) the alternative *fail to reject H₀*.

The next example is more extensive and summarizes the essence of many practical analyses. Its use can be extended to more variables, time series of data (Fig. 5.2).

Example We use De Souza-Carvalho's Wolfram demonstration "Correlation and Covariance of Random Discrete Signals" http://demonstrations.wolfram.com/ CorrelationAndCovarianceOfRandomDiscreteSignals/ to depict, Fig. 5.3, how two

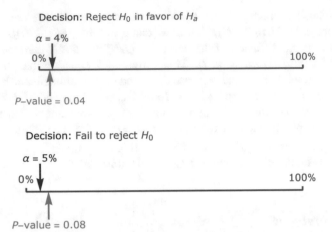

Fig. 5.2 Rejection and acceptance decisions for one-sided statistical hypotheses, the null hypothesis, H_0, and its alternative H_a using E Schultz Demonstration http://demonstrations.wolfram.com/DecisionsBasedOnPValuesAndSignificanceLevels/

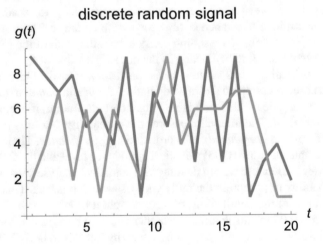

Fig. 5.3 Signals over time, t, for two time series of hypothetical data ($n = 20$ for both). Their study is exemplified using D de Souza-Carvalho http://demonstration.wolfram.com/CorrelationAndCovarianceOfRandomDiscreteSignals. The individual signals are joined by straight lines for ease of interpretation. It does not follow that actual (and omitted or otherwise unavailable) data on each segment are as depicted. They may fluctuate in unknown ways

sets of data track one another over time (there are two time series, one for x and one for y: $x(t)$ and $y(t)$). The demonstration includes the key numerical formulae used in the analysis. The "hat," ^, over the symbol indicates that the formula is an estimator of the numerical value obtained from the data. For instance, σ_{xy} is *covariance* between X and Y; it is numerically estimated as 0.232 and is symbolized by $\hat{\sigma}_{xy}$. The data consist of two samples (of equal size, $n_x = n_y = 20$), in the left panel. The shape of these two signals is depicted in figure to the right, which includes the formulae for the terms used in calculating the sample correlation coefficient and covariance. The two random sets of empirical data, signals, in the left panel appear to be uncorrelated: this is confirmed by the low estimated coefficient of correlation ($\hat{\rho}$ xy), which equals 0.034. The two sets of data are drawn at random with the following summary data:

- $\bar{x}$ = sample arithmetic average for X, 5.6
- $\bar{y}$ = sample arithmetic average for Y, 4.55
- σ_{xy} = covariance between the two variables, 0.232

In a set of data, either as cross-sectional samples or as one or more time series, the fact that the data for X and for Y may yield a large numerical coefficient of correlation, for instance, 0.85, does not mean that these two variables theoretically depend on each other although empirically they do. We note that the coefficient of correlation is a random variable (taking on values in the interval − 1, 1, and dimensionless) and thus an estimate of the coefficient has its own confidence intervals. An assertion of empirical dependence, particularly when used in the analysis of cause and effect, requires a theoretical basis that the empirical results either confirm or disconfirm. The issue is illustrated using the Anscombe *quartet*'s data and use these for estimating the linear relationship between them.

Example The four empirical relationships in Anscombe's four sets of sample data, drawn from the random variables X and Y, are developed using I Mac Leod Mathematica demonstration, http://demonstrations.wolfram.com/Anscombe Quartet/. The estimated coefficients of correlations ($\hat{\rho}_{xy}$, are, from left to right, *0.817, 0.816, 0.816*, and *0.816*); the linear equation $Y = 3.0 + 0.5X$ does not change regardless of the shape of the data. The demonstration also illustrates that the linear equation extends beyond the data. Attempting to either interpolate or extrapolate, for at least three data sets, is prima facie incorrect. The first panel provides a plausible result for both interpolation and extrapolation, at least within the confines of the sample. The other three panels below, Fig. 5.4, depict the issue that fitting a linear equation to data that have different configurations requires additional insights. These concern the nature of the data, the mechanism that links the two (random) variables, the theoretical basis for the mechanism, as well as several statistical question having to do with estimating the coefficients of the linear equation (3 and 0.50).

Correlations, when dealing with distributions of random variables, affect the output of simple operations between the distributions. For example, the multiplication of the distribution of two random variables is affected by their correlation. Figure 5.5 depicts two hypothetical alternative distributions of concentrations generated with and without correlations as the output of a hypothetical prediction of the distribution of a particular risk factor at a specific point and time.

The effect of the correlation is that the probability to the left of the regulatory standard (the areas under the two curves bounded by the regulatory standard) is much higher than it would be without accounting for the correlation. Visually, it is

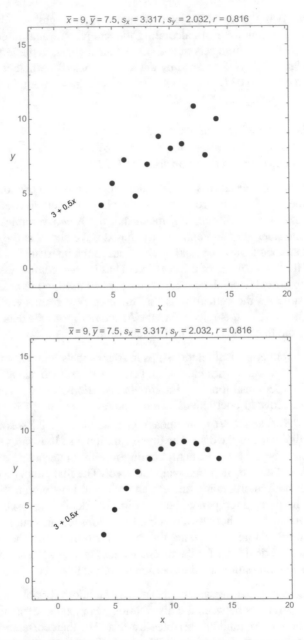

Fig. 5.4 Four possible but very different data sets result in the same linear equation developed by I I Mac Leod http://demonstrations.wolfram.com/AnscombeQuartet/

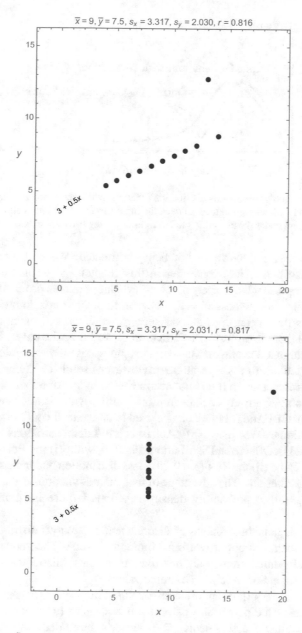

Fig. 5.4 (continued)

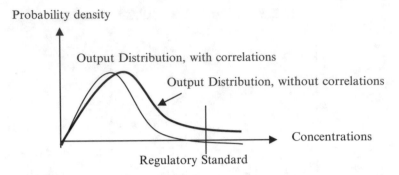

Fig. 5.5 Two continuous probability distribution functions of concentrations at a specific location and time with and without correlations between the input variables and implications regarding the areas (probabilities) under the two curves to the right of the regulatory standard

approximately twice the size of not being correlated. We will discuss and use probability distribution functions later in this chapter and throughout this book. Distributions represent the entirety of the knowledge about the probabilistic descriptions of the outcomes associated with an event. In other words, knowing the distribution implies that mean, median, variance, percentiles, and other quantities are fully known – there is no loss of information as would be the case when using the mean (e.g., arithmetic average). The average aggregates all the individual values of a random variable into a single number. However useful this number may be, it implies loss of information if it is not accompanied by measures of variability.

RJ Brown's Wolfram demonstration, http://demonstrations.wolfram.com/ ConnectingTheCDFAndThePDF/ can be used to understand continuous probability density functions. We use NM Abbassi http//:demonstrations.wolfram.com/ IllustratingTheUseOfDiscreteDistribution, Fig. 5.6, with: (i) the Poisson mass distribution ($\lambda = 2$)and (ii) the Zipf ($p = 0.25$) mass distribution and include their cumulative mass distributions. The advantage of using mass functions is that they directly provide probabilities; probability density functions require integration to obtain probabilities.

Figure 5.7 depicts three shapes of discrete and continuous distributions using J Bolte's http://demonstrations.wolfram.com/SampleVersusTheoreticalDistribution/ with statistically fitted continuous distributions to their histograms. The distributions are normal, exponential, and extreme values.

Continuous distributions are often encountered in assessing catastrophic events; measures such as the percentiles of probability density functions can be used to develop probabilistic assessments. C Boucher's http://demonstrations.wolfram. com/PercentilesOfCertainProbabilityDistributions/ allows modeling the percentiles for the standardized normal distribution, N(0, 1), and the 95th percentile encountered in many statistical discussions, including some of ours, Fig. 5.8.

Figure 5.9 depicts two different modes for levee failure, for example, poor maintenance and erosion, represented by individual cumulative distribution functions, combined into an aggregate cumulative distribution function (CDF). The

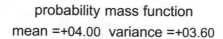

probability mass function

mean =+04.00 variance =+03.60

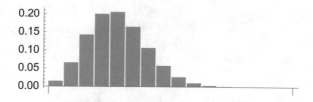

cumulative distribution function

quantile =   07.00
skewness = +00.42 kurtosis = +03.13

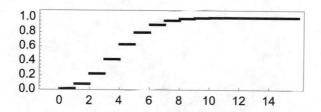

probability mass function

mean =☐   variance =☐

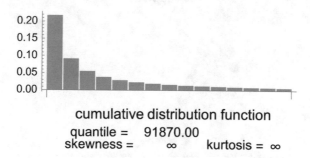

cumulative distribution function

quantile =   91870.00
skewness =   ∞   kurtosis = ∞

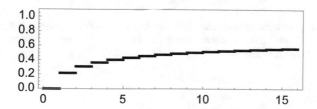

Fig. 5.6 Mass distribution functions. Discrete binomial mass distribution and its cumulative mass distribution for $n = 40$ and $p = 0.1$. Zipf mass distribution and its cumulative mass distribution developed with NM Abbassi http//:demonstrations.wolfram.com/IllustratingTheUseOfDiscrete Distribution

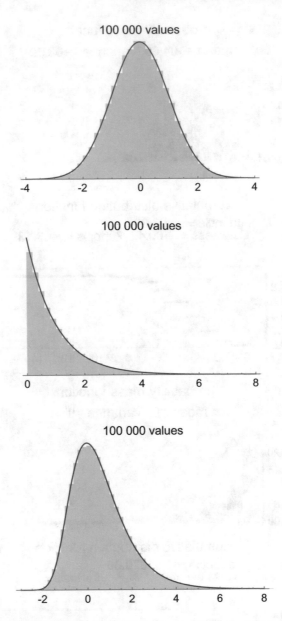

Fig. 5.7 Normal, exponential, and extreme values discrete and fitted continuous distribution functions to the discrete data obtained using J Bolte's http://demonstrations.wolfram.com/SampleVersusTheoreticalDistribution/

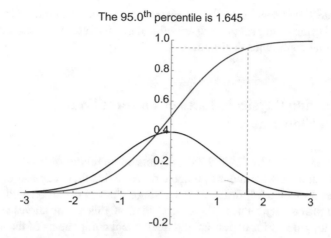

Fig. 5.8 Standardized normal and its cumulative distribution functions obtained using C Boucher's http://demonstrations.wolfram.com/PercentilesOfCertainProbabilityDistributions/

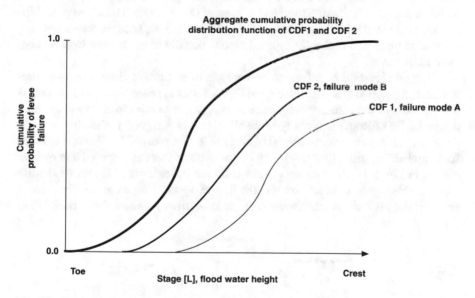

Fig. 5.9 Continuous cumulative distribution functions, CDFs, depicting two independent failure modes, A and B, and the aggregation of these distributions into a single CDF

mathematical operation that yields the aggregate CDF may be a numerical simulation that we depict with a continuous (and heavier line) line for simplicity's sake.

Numerical methods based on Monte Carlo (MC) simulations generally yield results such as those depicted in the previous example of discrete distribution functions. Those are discrete outputs to which continuous functions can be fitted, as discussed and depicted in the example. The MC distributions will be discrete, but, because the number of simulations is generally very large (e.g., as we will discuss in the Monte Carlo section of this chapter, we have used approximately 10,000 independent samples to generate a distribution), the discrete curves closely approximate the actual continuous distribution.

5.2   Modeling Physical Modes of Failure: Examples from Flow Rates

Figure 5.10, motivated by NAS (2000), depicts some relationships that arise in the context of flooding events and the probable damage that they may cause. For example, in panel 3 of Fig. 5.4, different water stage elevations can be associated with exceedance probability values. The central idea of this set of curves is a simple, generalizable, process for understanding the critical components of the analyses of probable catastrophic events: namely, the process *event → probability → failure → probability → damage*. For example, for *levees*, the probable non-failure point, *PNP*, is assigned pr = 0.15 and the probable failure point, *PFP*, pr = 0.85, each of which is determined through technical judgment (NAS 2000). These two probabilities are associated with the two heights of the flood stages; because levees are long structures, these two probabilities are defined spatially at the known weakest segment of the levee.

These relationships contribute to the analysis of alternative choices that we summarize in a decision analytic framework, Fig. 5.5, that is probabilistic and generalizable to most of the catastrophic incidents we discuss in this Book. We note that a deterministic outcome is actually probabilistic: it is assigned probability = 1. As discussed, its nonoccurrence has probability = 0. The prospective event, $e \in E$ has some probability pr(e), $[0 \le pr(e) \le 1]$; e cannot be further decomposed into smaller events. In Fig. 5.11, the data are identified by not to scale-small circles to identify them. In this figure, from left to right, the first two panels depict an idealized set of probabilistic choices (alternative choices emanate from a chance node, depicted as

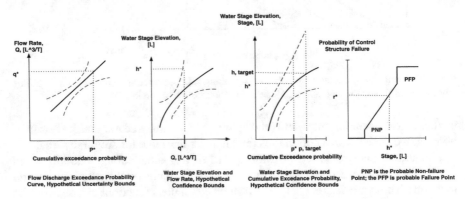

Fig. 5.10 Relationships between exceedance probabilities and structural failure PNP is Probable Nonfailure Point and PFP Probable Failure Point; L is length, T is time, p is probability

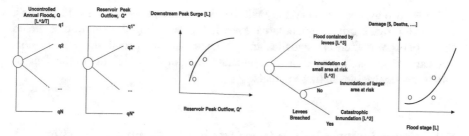

Fig. 5.11 Events and choices: probabilities and causation in a nutshell; data are depicted by small circles for ease of representations and limited to keep the diagram legible

large circles). Each chance node is mass distribution functions because each branch is probabilistic, as we have shown and discussed elsewhere in this book. Modeling and parameter estimation yield the nonlinear functions depicted in the third and in the last panel of this figure; the forecast is well beyond the historical data.

The choice of models is made by analyst based on her knowledge about the process she is describing. Consider the normal distribution: it concentrates that most of the data (95%) within ±3 standard deviations (*sd*) from the mean, mode, and median. The distribution is symmetric about these, unimodal and thin tailed. The rare events are in the tails of the distribution: those can be > or > > than 3sd or <, << than 3sd. Events that fall in these tails may have never been observed. Their nature may be completely understood. For example, an extreme event with probability $1*10^{-7}$ may not have an historical basis: there may be no known data although a theoretical understanding of the generating mechanism must exist otherwise the curve would not be correctly used. An incorrect prediction may result in serious public health consequences, misallocation of budgets, and have other consequences.

A critical concept in these discussions is the *annual exceedance probability* (EAP) of a particular value, such as a peak stream flood, that may result in a catastrophic inundation. A *very low* AEP is defined as <0.001; this is a percentile probability value of the distribution of peak floods (Asquith et al., 2017). Asquith et al. (2017, p. 11; parenthetic added) also define two important quantities to characterize uncertainty:

"1. *Distribution Choice Uncertainty* (σ_{dc})—This uncertainty is ... the standard deviation of the quantile estimates of all distributions considered. Its value is dependent on which distributions are chosen... . A different set or ensemble of distributions would yield a different distribution choice uncertainty. ...

2. *Sampling Uncertainty* (σ_s)— ... uncertainty of a given quantile estimate for a specific fitted distribution. For example, ... the sampling uncertainty for a quantile under the condition (assumption) that the chosen distribution is actually correct using the available sample data. This uncertainty can be constructed from analytical results, Monte Carlo simulation, or hybrid techniques... . The foundation of sampling uncertainty is ... the sampling variance-covariance (matrix) of either the moments (product or L-moment) and (or) the parameters of the distribution based on the data ... . The sampling variance-covariance matrix defines the coupling of variation among the ... parameters (and) provides a key to computing sampling uncertainty and also confidence limits for a given quantile."

These two aspects of uncertainty will be used in the paragraphs that follow as a means to provide critical quantitative knowledge regarding the single or coupled behavior of parameters of a statistical model. These, although not the only parameters that can be estimated, provide a good understanding of the variability of the data for one or more variables in those models.

5.3   Statistical Modeling: Brief Review to Complete Earlier Discussions

The importance of rare or extreme events, and the magnitude of their consequences, may elude most stakeholders in part because those events may not have been observed. The simplicity of the normal distribution, $N(.)$, appears to be based on its symmetry about its central moments (i.e., mean, mode, median), and quickly vanishing left and right tails. It is associated with the central limit theorem (CLT): a theoretical foundation in which the means of sample distributions converge, as the number of samples tends to infinity, to be normally distributed. The two tails are equidistant (e.g., > or > > than ±3 standard deviations) from the mean (the expected value) and may be negligible because the proportion of very small and very large values is much smaller than the total. This distribution, if X is the normally distributed random variable, $X \sim N(.)$, has the probability density depicted in the left panel of Fig. 5.12. When $X \sim N(.)$, the values of X, $x \in X$ with $-\infty < x < \infty$, μ is the central location parameter (the mean of the pdf), and $\sigma > 0$ is the scale parameter. The cumulative distribution function is depicted in the right panel of Fig. 5.12. The formula for normal probability density function, which is symmetric about its mean, mode, and median, is

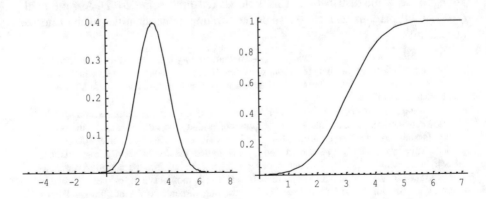

Fig. 5.12 Normal distribution of an arbitrary random variable with mean 3 and variance 1, $X \sim N(3, 1)$, first panel; the second panel depicts its cumulative distribution. The abscissa has arbitrary units; the ordinate is density and cumulative density, respectively

$$N\left(x;\mu,\sigma\right)=\left(\sigma\left(2\pi\right)^{\frac{1}{2}}\right)^{-1}Exp\left(-\left(x-\mu\right)^{2}/2\sigma^{2}\right).\qquad(5.2)$$

An example for $X \sim N(3, 1)$ and its cumulative distribution functions are depicted in Fig. 5.12.

I McLeod http://demonstrations.wolfram.com/StandardNormalDistribution Areas/ shows the implications of using probability density functions in the calculation of probabilities. Figure 5.13 depicts two areas under each of the tails: their z values are -1.65 and 1.65. If the sum of these two areas is subtracted from 1.00, the area under the entire curve, the result is the area in the center of the distribution. We also depict a single area, in which the z-value is 2.15.

The standardized normal distribution, in which the area under the entire distribution equals 1.00, is found as follows. Let $y = (x-\mu)/\sigma$, that is, $x = \mu + y\sigma$. For a continuous distribution the *probability* is the product of $f(x)$ and dx; that is, $pr(x) = f(x)$ dx. By substitution we obtain: $pr(x) = 1/(\sigma\sqrt{2\pi})[exp(-1/2y^2)d(\mu + y\sigma)]$. To find the density function of this random variable, let $d(\mu + y\sigma) = 0 + \sigma dy$. Then, we obtain $pr(x) = 1/(\sqrt{2\pi})exp(-1/2y^2)dy$. The standardized normal distribution of the random variable $Z \sim N(0, 1)$ is $f(z) = (2\pi)^{1/2} exp(-z^2/2)$, with $-\infty < z < \infty$. Its cumulative density function is $\Phi(z)=1/\sqrt{2\pi}\int_{\infty}^{z}exp[-(y^2/2)]dy$. The probability that the random variable Z is between the limits a and b $(a < b)$ is $pr(a < Z < b) = [\Phi(b)-\Phi(a)]$. The sum of two or more normally distributed random variables is also normally distributed, if those random variables are statistically independent. When the samples are smaller that *30*, the *t*-distribution instead. However, this is only a rule of thumb.

Asquith et al. (2017) tabulate nine distributions (including the Weibull 3 parameters and Pearson Type III distributions) that can be used to study AEPs in the context of different physical hazards. These authors include a high-level decision tree to help select between them: their work is of interest to those seeking further guidance in the analysis of AEP, particularly regarding site-specific stream flow analyses. We can consider a sample of data from a time series of events, either historical or systematically observed by metering flow rates in a channel of determinable slope and other hydraulic factors. Our focus shifts to its maxima and minima, rather the entirety of the data, over a period of record. What we seek to determine are periods of inundations or droughts: the distribution of the daily maxima or minima flow rates as extreme values. The data consist of the available historical record – for a specific water basin – from which several time series of flow rates, precipitation, and so on can be available from different sources. This record can be *broken, incomplete*: the former is due to random events such as the failure of a recording instrument, the latter by floods are too high to be recorded. Additional considerations include missing events, the effect of seasonal or climatic cycles; years with zero flood have to be accounted by special methods, as should outliers. Taking a hypothetical time series of maximal flow rates, Q; for a river basin we may consider the following critical quantities. An extreme event is an event that equals or exceeds a

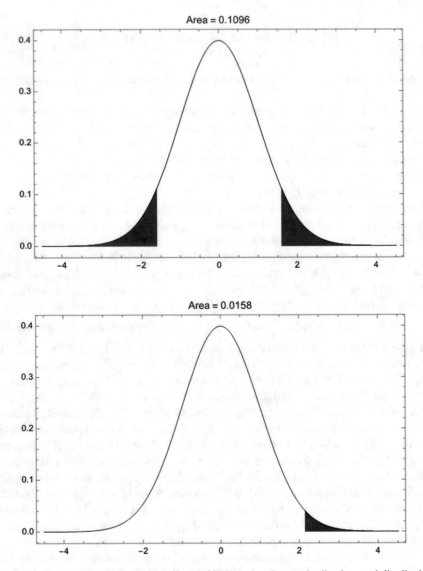

Fig. 5.13 Two aspects of calculating tail probabilities using the standardized normal distribution, the random variable $Z \sim N(0, 1)$; $z \in Z$. The y-axis measures the density of the distribution; the z-axis measures units of z. Developed using I McLeod http://demonstrations.wolfram.com/StandardNormalDistributionAreas/

specific magnitude; for example, flow rate $q*$, $q \in Q$. The choice of the magnitude for this threshold is a matter of engineering judgment. A threshold is defined as a specific value q_T, at time t, such that $q_T \leq q*$. A recurrence interval, τ, is the number of times that $q* \geq q_T$, in the record. The expected return period, $E(\tau)$, is the average recurrence interval between events equaling or exceeding the threshold.

Example We let $pr(q \geq q_T) = pr$. Using the binomial probability distribution, $pr(q \geq q_T$ at least once in n years). That is, $1 - (1 - pr)^n$; $pr(q \geq q_T) = 1/T$, the reciprocal of time [time in appropriate units]. Assuming that $q_T = 3000$ m³/second, if there are 11 exceedances $n = 10$. The probability, pr, that a flow rate ≥ 3000 m³/second occurs at least once in the next 10 years is approximately 0.096, that is, $1 - (1-1/100)^{10}$. The return period T [years] is the number of times the threshold value is either equaled or exceeded: pr = 1/T. The expected annual damage, EAD [$], is calculated over all probabilities of exceedance: $EAD = \sum_i (pr_i * D_i)$. The threshold is assumed to be constant; an assumption depends on changes to the stream bed, maintenance, and several other hydraulic factors.

The modeling of the magnitude of the event causing damage and the magnitude of the damage are described and discussed by the USACoE (1993) (Chap. 3, Figures on page 7–2) and the NAS (2000). We depict critical relationships between event and effect in Fig. 5.14. It is an epistemic approach, as evident from the data (the enlarged circles) and the approximate confidence bounds about those curves depicted in Fig. 5.14.

An important relationship is that between the probability of exceeding specific stage heights and the total damage that the associated floods (or other catastrophic incident) may cause in a year, Fig. 5.15. The total *expected annual damage*:

$EAD = \int_0^1 D(p)\,dp$, where $p =$ probability. Damage is heterogeneous: from ecological to human deaths; the unit of value is monetary, when it can be monetized.

Figure 5.16 depicts the relationship between the probability of exceedances and the magnitude of flow rates. The uncertainty is associated with the estimation of the location, and shape of the curve, conditioned on sampled data, is depicted by the dasOdot upper and lower confidence bounds on the average curve. In general, the probabilistic factor of safety using the design flow rate $q*$ corresponds to the upper 0.95 probability, namely, $q_{0.95}$, on q_{100}, the latter is the flow rate with a return period of 100 years (NAS, 2000). The expression for the upper and lower bounds on the $\log_{10}$(flow rate, q) is $\log(q) = <\log(q)> \pm k * \sigma_{\log q}$, where $<.>$ is shorthand for *expected value*, σ is *standard deviation*, and k is the number of standard deviations from the mean.

When analyzing two or more random variables, the computation of the expected values may involve addition and subtraction, for example, $\mu_z = \mu_X \pm \mu_Y$ yields the combined mean. However, the combined variance should account for their correlation, through the coefficient of correlation between X and Y, $\rho : \sigma_z^2 = \sigma_X^2 + \sigma_Y^2 \pm 2\rho_{XY}\sigma_X\sigma_Y$; hence, the combined *standard deviation* is $\sqrt[2]{\sigma_z^2}$. These formulae apply to population parameters. When using samples of data, these parameters are substituted by their sampling estimates: sample means, sample standard deviation, and sample coefficient of correlation. For two random variables X and Y, the linear coefficient of correlation is $\rho = cov(X, Y)/var(X)var(Y)$ or $cov(X,Y)/\sigma_X\sigma_Y$; $(-1 \leq \rho \leq 1)$, a dimensionless number. The sample *coefficient of correlation* (or *determination*, if squared) is a common statistic estimated using the formula:

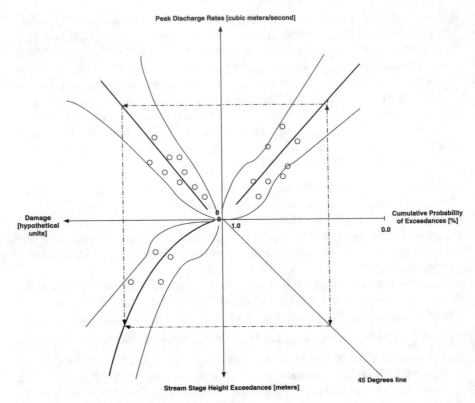

Fig. 5.14 Dependencies in modeling relationships between flood exceedance probability and damage. 1. All relationships are simplified to linear. 2. Data points enlarged for ease of reading. 3. Physical bounds affecting estimate ignored. 4. Note that the probability axis begins at 1.0, rather than the usual 0.0 value. 5. Upper and lower uncertainty bounds are indicative only and bulge to emphasize uncertainty at very low quantities or probabilities

$$R = \left(1/(n-1)\right) \sum_{i=1}^{n} \left\{ \left[(x_i - \overline{x})/s_X \right] \left[(y_i - \overline{y})/s_Y \right] \right\}. \tag{5.3}$$

In this formula, n is the sample size, $\overline{x}$ and $\overline{y}$ are the sample means, s_X and s_y are the sample standard deviations. When there are several independent variables, as would be the case when X, Z, W, and other variables are used in a model, the coefficient of correlation is called the coefficient of multiple correlations. The coefficient of correlation, as discussed, is a random variable with a specific distribution (e.g., it follows the t-distribution for a sufficiently large sample and under the null hypothesis $\rho = 0$). Statistical significance and the upper and lower confidence limits apply in the multiple variables case. We exemplify some of the computations involving random variables means and variances (or the squared root of the variances, the standard deviation) next.

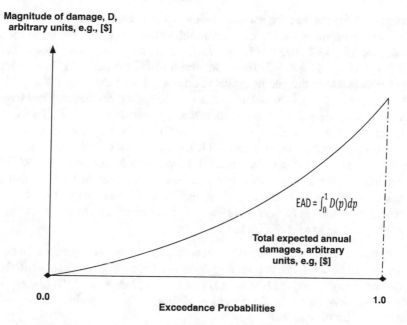

Fig. 5.15 Exceedance probabilities and total expected annual damage, arbitrary numbers

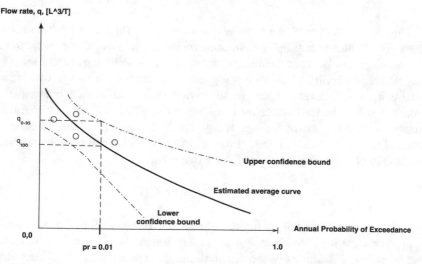

Fig. 5.16 Average peak flow rate curve and arbitrary uncertainty bounds (e.g., 5% lower and 95% upper bounds), for a period of N years. The specific flow rate associated with the annual exceedance probability = 0.01, q_{100}, has a 0.01 probability of being equaled or exceeded. The points are represented by circles for ease of depiction and their number is limited for the same reason. *Notes*: Annual exceedance probability: 1-(1-PR)^N, N = number of years. pr = exceedance probability, (1-pr) = non-exceedance probability. Example: Return period (T = years): therefore annual exceedance pr = 1/100 = 0.01. Annual non-exceedance probability = 1–0.01 = 0.99. Upper and lower bounds are determined by the chosen confidence level, a probability, and specific formulae. Data from a floor frequency analysis given an hypothetical N years record of flow rates. Not all data shown

Example Suppose that the sample means of three random variables are $X = 10$, $W = 5$ and $Z = 1$; the units, [.], of these variables are unstated but should be the same. Then, $ln(10) = 2.302$, $ln(5) = 1.61$, and $ln(2) = 0.69$: their sum is $2.302 + 1.61 + 0.69 = 4.602$. The geometric mean of the output, Y, is calculated as $exp(4.602) = 93.68$; the geometric standard deviation of the output Y is calculated as: $s_{geometric} = exp(\sqrt{\sigma^2})$, in which σ^2 is the variance of the *log*-transformed data. The three variances are 3.5, 1.30, and 0.10, respectively, and sum to 4.90. Therefore, the geometric standard deviation of Y is $exp(\sqrt{4.90}) = 9.14$. Suppose we wish to combine the 95% upper confidence limits (UCLs) of variables in a formula; these limits *are* random variables. Such combination, however, does not produce a 95% confidence interval. For instance, let X and Y be independent standardized normal distributions, $\sim N(0, 1)$, respectively, then the sum of their 95% UCLs, namely, $1.96 + 1.96 = 3.92$, is $(3.92/1.4142) = 2.77$ standard deviations above the mean of 0, corresponding to a 99.4% UCL.

Suppose a physical system has two mechanisms of failure over time. In terms of distribution functions, one is normally distributed, the other is an exponential distribution of failures: $F_1(t) = [1-N(\mu_T, \sigma_T)]$ and $F_2(t) = [1-exp(-\lambda t)]$. The superposition of the *complements* of the distributions of time-to-failure yields the survivorship distribution for the system:

$$F_{1,2}(t) = \left[1 - N\left[\left(\mu_T, \sigma_T\right)\right]\right]\left[1 - exp\left(-\lambda t\right)\right].$$ (5.4)

This result also applies to a *parallel* system with two (or more) components with known (or assumed) distributions. In some instances, the analyst may want to indicate her degree of belief in a choice of distributions by weighing each of them by a judgmental probability. Consider heterogeneous failure rates of two groups and suppose that those failures are due to some mechanistic difference between those groups. Let $F_1(t)$ be the cumulative distribution of failures in group *1* and $F_2(t)$ the cumulative distribution of failures for group *2*. The *superposition* of these two cumulative distributions of failure times is their sum, which can be weighted by the probability of belonging to either of these two groups:

$$F_{1,2}(t) = pr_1 F_1(t) + pr_2 F_2(t),$$ (5.5)

where $pr_1 + pr_2 = 1.00$. For example, suppose that the probability that a proportion π_1 of more sensitive individuals is pr_1, and the probability of other proportion such that $(1-\pi_1) = pr_2$. Assume that the distributions are exponential with failure rates λ_1 and λ_2, respectively. The superposition of these two distributions is $F_{1,2}(t) = \{pr_1[1-exp(-\lambda_1 t)]\} + \{pr_2[1-exp(-\lambda_2 t)]\}$, with $\lambda_1 > \lambda_2$. Suppose a system has two mechanisms of failure over time; one is normally distributed, the other is exponentially distributed. That is $F_1(t) = [1-N(\mu_T, \sigma_T)]$ and $F_2(t) = [1-exp(-\lambda t)]$. The superposition of the *complements* of the distributions of time-to-failure yields the survivorship distribution for the entire system: $F_{1,2}(t) = [1-N[(\mu_T, \sigma_T)][1-exp(-\lambda t)]$. In some instances, the analyst may want to indicate her degree of belief in a choice of distributions by weighing each of them by a judgmental probability. Consider heterogeneous failure rates of two groups in a single population and suppose that those failures are due to their genetic differences. Let $F_1(t)$ be the cumulative distribution of failures in group *1* and $F_2(t)$ the cumulative distribution of failures for group *2*. The superposition of these two cumulative distributions of failure times is their sum, which can be weighted by the probability of belonging to either of these two groups: $F_{1,2}(t) = pr_1 F_1(t) + pr_2 F_2(t)$, where $pr_1 + pr_2 = 1.00$.

Physical mechanisms can be modeled through algorithms that yield numerical simulations of the trajectories of the physical hazard (e.g., a hurricane). Different predicted solutions generate an ensemble of predictions. Those solutions can be spatially distributed as depicted in Fig. 5.17: the trajectories are plotted as they cross the geographic boundary D-D. It can be a natural boundary such as the physical border between land and sea. For example, 95% of the predictions from different models cross that boundary. If the boundary is shifted – changing its latitude and longitude – then the trajectories may diverge more significantly: only a few may fall within that interval.

The reason for the differences in the trajectories can be due to the assumptions made.

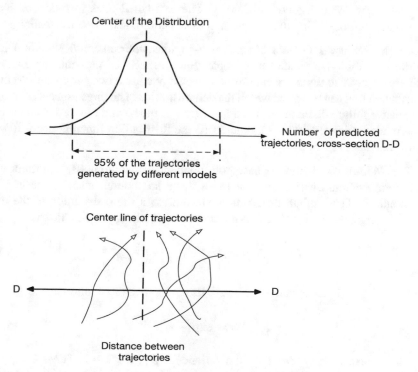

Fig. 5.17 Combination of trajectories and their distribution at the boundary D-D; only a limited number of trajectories are depicted; the actual number can be much larger. Trajectories outside the 95% confidence limits may or may not be omitted from an actual analysis

Regarding the choice of a distribution, the combination of practical and theoretical considerations suggests the specific distributions that may be used to describe the probabilities associated with the consequences from a catastrophic event. If the data are continuous, we will either write *probability density function* or the more general phrase *distribution function* when the context is clear. If the data is discrete, we will write *probability mass function*, but may also use *distribution function*. Examples of distribution functions with thin tails are the normal and the negative exponential distributions. What conceptually separates rare from routine events is the thickness of the tail or tails of the distribution: the distribution used has a single long right tail. It covers the domain from large to extreme magnitudes. The probability density function of a fat-tailed distribution, for large $x \in X$, goes to zero as $x^{-\alpha}$: fat-tailed distributions are long-tailed. Some distributions have a tail which goes to zero slower than an exponential function (meaning they are heavy-tailed), but faster than the $x^{-\alpha}$ (meaning they are not fat-tailed), for example, the log-normal distribution. Other distributions, example, the Pareto distribution, are also fat-tailed. The statistic of importance for these determinations is the moment generating function of the distribution, not discussed for brevity's sake. The probability that some large magnitude in the tail of the distribution will exceed any other previous level approaches 1. Single right-tailed distributions that have fat tails include Pareto, log-normal, Lévy, Weibull, and log-Cauchy. The two-tailed ones include Cauchy's, stable distributions (other than the normal distribution), and the t-distribution.

Example We use J O'Hara http://demonstrations.wolfram.com/ReliabilityDistributions/ReliabilityDistributions/ to depict three aspects of probabilistic analysis that are essential to assessing catastrophic events. We use a two parameters Weibull distribution that can be varied within the demonstration. The three aspects are the (i) cumulative failure distribution functions, $F(t)$; (ii) the hazard rate or instantaneous failure rate, $h(t) = (f(t))/(1 - F(t))$; and (iii) the distribution function itself, f(t), in which t is time, Fig. 5.18.

The Weibull distribution is based on independent, continuous, minimum values from a parent distribution that has a finite left bound (such as the *gamma* distribution). The Weibull distribution can represent the distribution of the lifetimes of physical objects. The *two*-parameter Weibull density function is.

$$f(x) = (\alpha/\beta)(x/\beta)^{\alpha-1} \exp\left[-(x/\beta)^{\alpha}\right]$$ (5.12)

with $x \geq 0, \alpha > 0, \beta > 0$; otherwise it is 0. Its cumulative distribution is

$$F(x) = 1 - \exp(-x/\beta)^{\alpha}$$ (5.13)

The mean equals $\beta\Gamma(1/\alpha + 1)$; the variance is $\beta^2\{\Gamma[(\alpha/2 + 1)-\Gamma(1/\alpha + 1)]^2\}$. If $\alpha < 1$, the density function is shaped as a reverse J. When $\alpha = 1$, the Weibull density becomes the exponential distribution; when $\alpha > 1$, the density function is approximately *bell*-shaped and symmetric. The two-parameter Weibull distribution can

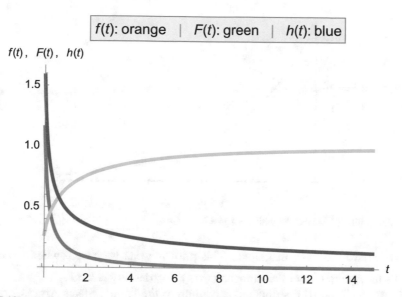

| $f(t)$: orange | $F(t)$: green | $h(t)$: blue |

Fig. 5.18 Distribution, cumulative distribution and hazard function for a two parameters Weibull distribution, Eq. 5.12 ($\alpha = 0.5$ and $\beta = 1.0$), obtained using J O'Hara http://demonstrations.wolfram.com/ReliabilityDistributions/ReliabilityDistributions/

account for constant, decreasing, and increasing failure rates. For $\alpha = 1, \beta = 3$, the two parameters Weibull distribution function plots are depicted in Fig. 5.19.

The exponential distribution, $f(t) = \lambda \text{Exp}(-\lambda t)$ can be used in Bayesian analysis, in which the *prior* distribution for λ is $\Gamma(a,b)$ and the posterior is $\Gamma(a + r, b + T)$. In these expressions, $\Gamma(.)$ is the incomplete gamma function. Suppose that a system can be repaired and that we are concerned with the *median time between failures*, MTBF, of its components. Other concepts, such as time to failure, can be analyzed with similar expressions. The estimated (indicated by the "hat") value of the median time between failures (MTBF) (for a credible interval,[1] $1 - \alpha$, is)

$$\widehat{MTBF} = \left(\Gamma^{-1}\left((1-\alpha), a', (1/b')\right)\right)^{-1} \tag{5.6}$$

In this equation, a' and b' are updated estimates obtained by maximizing their likelihoods using the maximum likelihood estimator. The historical data on the total number of failures (#F) and the total time under operation, T: a and b, respectively. If the system has not yet failed, or it is a new system entirely, the total number of likely failures can be approximated using expert advice. For example, consensus between experts can be formulated as prior knowledge stated as: (i) median MTBF; (ii) low MTBF; or (iii) as a *weak prior* MTBF. The latter of these three is based on the consensus MTBF, and on the assumption that $a = 1$ such that the mean and the

---

[1] The *credible interval* is the Bayesian equivalent of the traditional probabilistic *confidence interval* $(1 - \alpha)$.

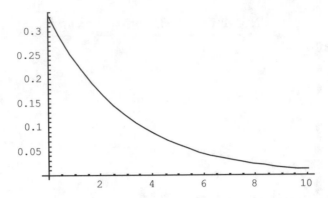

Fig. 5.19 Weibull Distribution with $\alpha = 1, \beta = 3$

variance have the same magnitude. This prior is weak because – when the mean equals the square root of the variance ($\sqrt{\sigma^2}$) – variability is large.

Pearson's Type III distribution, according to the United States Army Corps of Engineers (USACoE 1993), states that the frequency distribution of annual maximum stream flows rates, Q, $q \in Q$, is the Pearson's Type III probability density function (DOT/FAA, 2003):

$$f(q) = \frac{(q-\gamma)^{\alpha-1}}{\beta^\alpha \Gamma(\alpha)} \text{Exp}\left(\frac{-(q-\gamma)}{\beta}\right) \qquad (5.14)$$

Where $\Gamma(.)$ is the gamma function. Its parameters are the mean (magnitude), variance (the slope of the distribution function), and skew (the curvature), which are estimated by the sample mean, variance, and skew.

The distribution of extreme events, $M_n = \max\{x_1, ..., x_N\}$, completes this discussion. These extremes are observations that can be associated with events over the years for which records exist. If the observations are independent and identically distributed (i.i.d.), their distribution converges (as the sample size $n \to \infty$) to the generalized extreme value distribution, GEV, $G(x)$:

$$G(x) = \text{Exp}\left(-\left(1+\xi\left(\frac{x-\mu}{\sigma}\right)\right)^{-\frac{1}{\xi}}\right) \qquad (5.15)$$

Where ξ is the shape parameter; μ, the location parameter; and σ, the scale parameter, which are estimated from the maxima of a sample of sufficiently large size. The estimates of these three parameters can be obtained using the MLE method or other method, such as the L-moments. Once these parameters are estimated, the GEV distribution is fully parameterized. For example, it can be used to calculate $\text{pr}(M(.) > z_m) = 1/m$: namely, the 1/m probability of exceeding the level z_m. This threshold (e.g., in units of flow rate [L^3/T]) is a constant, u, such that [Y = (X-u), given that X > u]. Assuming that the differences (x_i − u) are i.i.d, their distribution converges to the generalized Pareto distribution, GPD: $H(Y) = 1 - (1 + \xi/\sigma)^{-1/\xi}$, with shape parameter ξ and scale parameter σ. The estimation of the parameters of GEV and GPD distributions implies that each parameter also has an associated distribution function. Upper and lower confidence limits (e.g., $u_l < \xi < u_u$) are used in interpreting the significance of results obtained through statistical analysis.

Estimation depends on the data (e.g., combining the historical record with systematic observation, accounting for missing information, and so on). The approaches may be parametric or nonparametric. Parametric methods consist of selecting the appropriate distribution function, a set of data, and produce a fitted curve to it, according to a criterion such as minimizing the squared distance between the actual point and the fitted point. The choice of estimation method depends on the analysts' understanding of the data. Fitting a curve to the data is not the sole objective of estimation. Another key objective is forecasting from the sample data, given the function for which we have obtained estimates of its parameters. The operations associated with this effort include interpolation and extrapolation, accounting for the uncertainty (measured by confidence bounds about the central estimate of the curve). Estimation depends on the nature of the data (e.g., time series versus cross-sectional). For example, the ordinary least squares (OLS), the maximum likelihood estimator (MLE), probability weighted moments, and L-moments may be appropriate if their assumptions are met in practice. The L-method yields estimates that are (i) robust to outliers; and, (ii) small sample bias tends to be small. L-moment estimators can be used when the maximum likelihood estimates are unavailable or difficult to compute.

Given a sample of historical stream flow peak data for a specific basin, and after taking the logarithms (base 10) of these data, $q \in Q$, the sample mean is: $\bar{q} = 1/n \sum q_i$ and variance, $s^2 = \dfrac{\sum_i (q_i - \bar{q})^2}{N-1}$, the computational formula for G(q) is (USACoE, 1993, p. 3–2):

$$G = \frac{N^2 \left(\sum_i q_i^3\right) - 3N\left(\sum_i q_i\right)\left(\sum_i q_i^2\right) + 2\left(\sum_i q_i\right)^3}{N(N-1)(N-2)s^3}. \tag{5.16}$$

Exceedances are calculated as $\log(q) = \bar{q} + Ks$, where K is the Pearson's Type III *deviate* relating the probability of exceedances to G. For example (USACoE, 1993, p. 3–3; Table 3–1), the logarithms of the event magnitude corresponding to each of the selected percent chance exceedance values are calculated using $\log(q) = \bar{q} + K*s = 3.3684 + 2.8236*(0.2456)$, Q = 11,500 [cubic feet per second] (or 325.644 m^3/sec). K depends on the skewness and percent chance p (Appendix V-5, USACoE). Outliers – for example, floods above or below specific thresholds – must also be assessed because they affect the estimation of these three parameters (USACoE, p. 3–5). A *risk*, in these analyses, is the probability of exactly n floods, which is obtained using the binomial probability mass distribution:

$pr(n) = (N!/(n!(N-n)!))pr^n(1-pr)^{N-n}$. The probability, pr, of a flood of magnitude q is estimated from a known record of floods. The probability distribution of binomial outcomes (i.e., an event can be either *flood* or *no flood*) accounts for a specific sequence of events over N years that could have occurred over that period of years. It follows that the probability of zero floods is $pr(0) = (1 - pr)^N$ and the probability of 1 or more floods is $1 - (1 - pr)^N$. For example, assuming that the probability of an event q is 0.01; then the *risk* of being flooded over a period $N = 30$ years is $1 - (1-0.01)^{30} = 0.25$ ($0 \le$ risk ≤ 1).

An aspect of the information that can be used to assess the probability of failure throughout the life of a unit at risk can be analyzed using the Mathematica demonstration by KS Lowder http://demonstrations.wolfram.com/ClosedFormFullLife CycleDistribution/, not shown. The demonstration depicts full life-cycle distributions. It includes probability and cumulative density functions, and the hazard function of the profile failures over time: (i) failure associated with being a new unit, (ii) the effect of random events over time, and (iii) the terminal event ending the lifecycle. The analysis that can be conducted approximate events throughout the lifecycle including the *wear-out*. These are the effect of being vulnerable in the earliest stages of life, a tendency to gain strength in some intermediate time period, and the inevitable effect of aging on survivability. Figure 5.20 clarifies these concepts and adds information that can be useful to sketch alternatives the *bathtub* curve. Although we have used time as the random variable, the distribution can be generalized to other random variables such as those that have distance as the argument of the function. Alternatives to the bathtub distribution can be developed from empirical data.

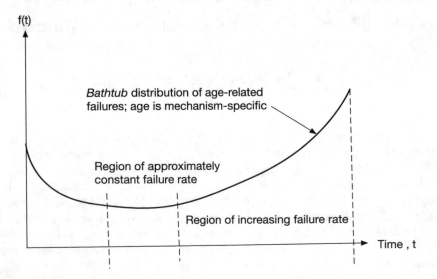

Fig. 5.20 Bathtub distribution of failure from ageing

In the previous discussions we have used analytical models. In many cases, their solutions are numerical rather than analytical. An approximation is preferred to an exact formula that, in many complicated cases, may not be achievable. We discuss and exemplify simple Monte Carlo simulations as a means to achieve those approximations.

5.3.1   Monte Carlo Simulations

Monte Carlo simulations can be exemplified using Crystal Ball™ as a means to obtain the distribution of the output variable for a set of *assumed* or *known* random variables for which the distributions are known either empirically or from theoretical reasoning. Figure 5.21 depicts the fit to the histogram generated by a Monte Carlo simulation using a *log*-normal distribution with mean *4.60* and standard deviation *0.36*, from the previous example. The continuous curve is a density function; density is measured on the *y*-axis.

Specifically, Monte Carlo simulations replace integration to provide numerical *approximations* of statistical quantities such as the mean, variance, other moments of the distribution and approximate probability distributions about these parameters. We extend the Monte Carlo simulation to a simple formula in which the inputs are random variable with known distributions.

Example   Consider the formula $(A*B*C*D)/(E*F)$. Let $A = 12$, $B = 50$, $C = 30$, $D = 0.75$, $E = 70$, and $F = 10$. All variables are uncorrelated and have arbitrary units. The deterministic result is $X = (12*50*30*0.75)/(70*10) = 19.286$. We change this determinist result by letting the six input variables be random with distributions: $A \sim N(12, 3.5), B \sim U(0, 100), C \sim U(1.00, 45), D \sim T(0, 0.75, 1.00), E \sim T(5, 70, 120)$, and $F \sim U(0, 20)$. Here, $A \sim N(12, 3.5)$ means that A is normally distributed with *mean 12* and *standard deviation = 3.5*; $B \sim U(0, 100)$ means that B is uniformly distributed with minimum value *0* and maximum value $= 100$; $D \sim T(.)$ means that D has triangular distribution with three values. The result of a Monte Carlo simulation with *10,000* trials yields the distribution of X, Fig. 5.22. The statistical results for X are *mean = 8.34, median = 0.95, standard deviation = 195.42*.

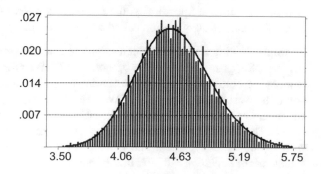

Fig. 5.21 Monte Carlo simulation and log-normal density function, LN(4.60, 0.36), fit to the simulation

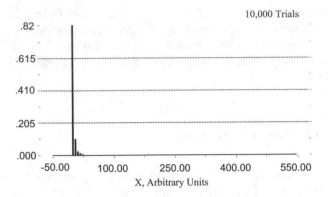

Fig. 5.22 Monte Carlo simulation results from Y = (A∗B∗C∗D)/(E∗F), all random variable in arbitrary units, after 10,000 independent trials

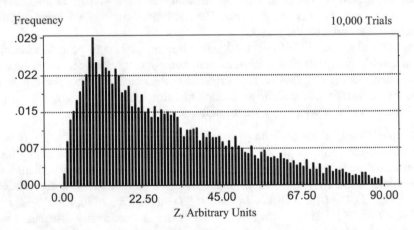

Fig. 5.23 Monte Carlo simulation of the ratio of random variables with distributions different from the normal or log-normal

What about ratios of random variables? If the random variables are independent and their *pdfs* are either normal or log-normal, the result is obtained by taking the ratio of the *pdfs*. However, this is not the case when the *pdfs* are triangular, uniform, or take other forms. Specifically, the *back-calculation* of probability distributions generally requires an operation called *deconvolution*.

Example Taking $Z = X*Y$, we can calculate Z as the product of two deterministic numbers. Suppose $X = 5$ and $Y = 5$, then $Z = 25$. We now change these variables to random variables with specific distributions. The question is: What is the distribution of the variable Z, for $Z = X*Y$? We use a Monte Carlo simulation to answer this question. Let X and Y be uniformly distributed: $X \sim U(1.00, 10.00)$ and $Y \sim U(1.00, 10.00)$. The empirical distribution of $Z = X*Y$, obtained by Monte Carlo simulation with *10,000* trials, Fig. 5.23. The statistics of Z are *mean = 30.09*, *median = 24.66*, and *standard deviation = 21.21*. In this example, we see that Z is not uniform.

The analyst will find situations in which the uncertainty in the values of a variable cannot be described by a distribution function; alternatively, an expert may be unwilling to provide such distribution. In these situation intervals, two ordered numbers can characterize uncertainty. The loss is that we do not know what happens between the limits of the interval or intervals. We might consider using the uniform distribution, which represents limited knowledge about the random variable. A justification for the uniform distribution is the *symmetry principle* whereby, when the outcomes are homogeneous and *nature* cannot influence one outcome over another, then each event has an equal probability of occurring. A second justification is based on the idea of entropy of information: the uniform distribution maximizes the uncertainty. However, it can be the case that these reasons are inappropriate because the expert does not have sufficient knowledge to assert either of them but, nonetheless, does so to be able to use Bayes' rule (which would yield a zero posterior distribution, if the expert asserted zero prior knowledge (but the likelihood is statistically significant). Of course, an analyst may also be unwilling to assert a distribution but may be willing to state lower and upper limits for that value. This form of inexact reasoning suggests plausible bounds on the upper and lower limits of an uncertain value using intervals.

Example The interval *[2, 15]* associated with the hypothetical variable AGE is understood to mean that an individual age can be either *2* or *15* and that nothing else is known about that variable. By comparison, consider the statement *AGE is uniformly distributed between 2 and 15 years.* This statement implies the uniform distribution, $U(a = 2, b = 15)$: each age ϵ AGE is equally likely. More generally, the uniform distribution $U(a, b)$ has a probability distribution $1/(b-a)$ for $a < x < b$, and zero otherwise. Its cumulative distribution is *0*, for $x \leq a$; $(x-a)/(b\ a)$ for $a < x < b$; and *1* for $x \geq b$. The expected value and the variance of the uniform distribution are $(a + b)/2$ and $(b-a)^2/12$, respectively.

This brief discussion serves as an introduction to the (deterministic) algebra of intervals. Table 5.1 depicts and exemplifies some of its operations.

Logical operators such as *AND* as well as *OR* applied to events *A* or *B* result in *envelopes* of intervals: *(A AND B) = envelope[max(0, A + B-1), min(A, B)]*, and *(A OR B) = envelope[max(A,B), min(1, A + B)]*. Intervals correctly portray uncertainty about a quantity when the magnitude of that quantity *is* just an interval, rather than a distribution. Moreover, the dependence between the quantities represented by A and B do not have to be known because the resulting envelopes, although optimal, do not require that knowledge.[2] The practical importance of not having to make the assumption of independence has two aspects. First, it simplifies probabilistic calcu-

---

[2] Based on Frechet's inequalities in real analysis (Moore RA, *Methods and Applications of Interval Analysis*, SIAM, Philadelphia, PA (1979). Ferson S, Reliable Calculations of Probabilities: Accounting for small sample size and model uncertainty, in *Intelligent Systems: A semiotic perspective*, NIST, Oct. 1996; Ibid. *What Monte Carlo Methods Cannot Do, Human and Environmental Risk Assessment*, 2:990 (1996), stated these bounds as: *max[0, pr(E) + pr(F)-1] ≤ pr(E AND F) ≤ min(pr(E), pr(F)*, and *max[pr(E), pr(F)] ≤ pr(E OR F) ≤ min(1, pr(E) + pr(F)*.

Table 5.1 Operations and example of results obtained using intervals

Interval operation	Resulting interval	Example	Comment
[a, b] + [c, d]	[a + c, b + d]	[23.54, 32.09] + [54.1, 98.5] = [77.64, 130.59]	None
[a, b]-[c, d]	[a-d, b-c]	[23.54, 32.09]-[54.1, 98.5] = [−74.96, −22.01]	None
[a, b]*[c, d]	Min[ac, ad, bc, bd], Max[ac, ad, bc, bd]	[23.54, 32.09]x[54.1, 98.5] = [1274.51, 316.87]	c, d > 0
[a, b]/[c, d]	Min[a/c, a/d, b/c, b/d], max[a/c, a/d, b/c, b/d]	[23.54, 32.09]/[54.1, 98.5] = [0.239, 0.593]	c, d > 0

lations: $pr(A \ AND \ B) = [pr(A)pr(B|A)]$ simplifies to $[pr(A)pr(B)]$. The last expression states that the probability of event A does not affect the probability of event B: A and B are stochastically independent. Second, we work with the marginal distributions of two (or more) random variables to *bound* the uncertainty of the dependent variable using operations (such as addition, multiplication, or other) performed on the two marginal distributions, even if there are undisclosed dependencies such as correlations.[3]

The example that follows includes deterministic, probabilistic, and interval analyses to complete our understanding of uncertain reasoning.

Example Suppose that exposure to an agent is stated to be the result of being exposed between *20* and *45* years, at concentrations between *0.01* and *10* micrograms/liter of water drunk, given that between *1* and *3* liters of water are drunk per day. The amount of mass ingested may be calculated using the simple formula *Exposure = [(mass input)/(unit of exposure)]*(period of exposure)* with the formula:

$$Exposure\left[Mass, micrograms \ of \ agent\right] = A\left[\mu g \ of \ agent \ / \ liter \ of \ water\right]$$
$$* B\left[liter \ of \ water \ ingested \ / \ day\right] * C\left[number \ of \ days\right]$$

Deterministic Analysis The deterministic calculation yields $E = A*B*C = 0.01*2*1600 = 32$ micrograms of the agent. The minimum is $(0.01)(7300)(1.00) = 734$ micrograms; the maximum is $(10)(16,425)(3) = 492,750$ micrograms.

Probabilistic Analysis Suppose that the probability distributions of the random variables A, B, and C are known to be: *log-normal$_A$ (0.01, 0.001)*, *uniform$_B$ (1, 3)*, and *triangular$_C$ (200, 1600, 3600)*. The mean of the uniform distribution is 2 liters per day and the standard deviation is *0.57* liters per day; the mean of the triangular

---

[3] Williamson RC, T Downs, Probabilistic Arithmetic I: numerical methods for calculating convolutions and dependency bounds, *Int. J. Approx. Reasoning* 4:89 (1990).

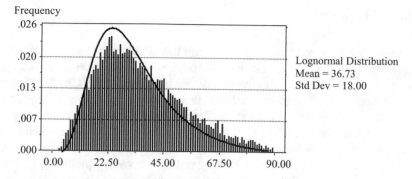

Fig. 5.24 Monte Carlo simulation log-normal$_A$ (0.01, 0.001), uniform$_B$ (1, 3) and triangular$_C$ (200, 1600, 3600). The result is LN(36.73, 18.00), fit to the empirical mass distribution simulation

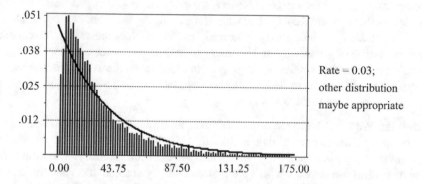

Fig. 5.25 Monte Carlo simulation and exponential distribution fit to the histogram ($\lambda = 0.03$)

distribution is *1600* days with standard deviation *725.72*. A Monte Carlo simulation (approximately *10,000* trials, assuming *independence* between the input variables) of the expression yields the following empirical distribution, to which we fit a *lognormal* distribution, Fig. 5.24.

The mean of the *empirical* distribution is *36.73* micrograms of the agent, the median is *32.97* micrograms, the standard deviation is *18.00* micrograms and the standard error of the mean of the sampling distribution is *0.18*. As a form of sensitivity analysis, we change the standard deviation of the random variable *A* to be *0.01* (the same value of the mean), with nothing else changed. We obtain the output distribution to which we have fit an exponential distribution, Fig. 5.25.

The mean of the empirical distribution is *36.26*, the median is *22.73*, the standard deviation equals *44.79*, and the standard error of the mean equals *0.45*. What is noticeable is that the tails of the new distribution are quite different from the baseline calculation.

We next use *intervals*. Suppose that the uncertainty in *A*, *B*, and *C* can only be represented by three intervals: *A = [0.00, 5]*, *B = [1, 3]*, and *C = [200, 3600]*. The uncertainty about the output variable, measured as an interval, is *[0.00, 54,000]*.

5.3.2   Comment

Intuitively, if a curve were to fit the data better than another, the *better fit* would support using the curve that best fits the data. The other would be discarded. Unfortunately, the best fit criterion can be wrong. If the best fitting curve were a mathematical representation of the theory that relates the input data to the depicted output data, or both sets of data were obtained through a competently conducted experiment, then the best fit criterion could be appropriate. The curve could be used for description and, in latter case, also for prediction. For rare or extreme events, this simplistic representation fails in part because the needed predictions – one or two points on this diagram – are orders of magnitude away from the median of the distribution. Historical data are essential to understanding the nature of information and the generating mechanisms of that information. That, to paraphrase a well-known saying, *those who neglect history are destined to repeat past mistakes*. The problem is that historical data suffer from issues ranging from not being available to having been falsified, incorrectly measured, and so on. Nonetheless, historical data, their resolution or granularity and behavior over time and space are essential for an equally extensive set of reasons, ranging from cross correlating different time series to generate mechanisms to describing the length of oscillations, transient phenomena, steady-state behavior, and so on.

Example   We use J Mcloone http://demonstrations.wolfram.com/DataSmoothing/ to exemplify a time series of data and a fitted continuous line to them. We select a data set (n = 300) in which the *y*-values and *x*-values are arbitrary. The time series depicts several oscillations, which we mathematically smooth and visually identify three as being remarkable after the smoothing process is applied. We also find that the overall trend is negative, Fig. 5.26.

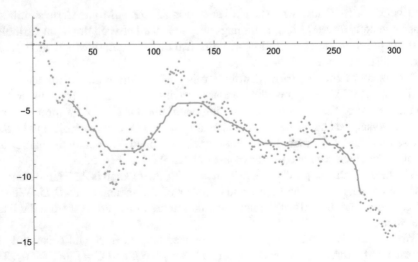

Fig. 5.26 Time-series analysis using a local smoothing estimator using http://demonstrations.wolfram.com/DataSmoothing/

Smoothing applied to data helps to identify peaks, valley, and local trends when the variability is complicated. If the data on the y-axis have physical or other meaning, such as economic or financial, the peaks, valleys, and local trends can be used to identify states of local disequilibrium and lead to other investigations. In continuous time or frequency (and for suitable spatial dimension such as distance from a known origin), these comments suggest using analyses based on waves or wave-like behaviors. The properties of a wave are often relevant to understand a number of physical phenomena that are associated with technological and natural catastrophic incidents. A Zheng http://demonstrations.wolfram.com/MovingWaveAnalysis/ can be used to understand the fundamentals of wave analyses, not shown. Regarding cyclical responses, often-encountered terms are period and frequency that characterize wave-like behaviors over time, space, or both, We use E Zeleny's http://demonstrations.wolfram.com/PeriodAndFrequency/ to model different waves. Fig. 5.27 depicts an initial frequency of 0.5 and each successive wave doubles that frequency which is the number of periods per unit time (i.e., second).

The spatial representation of two sources of waves, one with amplitude -7 and the other with amplitude 6, frequency 2.46, and phase $= 1.0$, is depicted using J Nochella http://demonstrations.wolfram.com/3DWaves, Fig. 5.28. One source of energy is roughly to the east and the other to the west of the surface.

5.4 Conclusion

In this chapter we have discussed and exemplified some of the critical elements of the quantitative analyses conducted in the context of catastrophic incidents, regardless of their nature. We have kept the technical details to the minimum that is con-

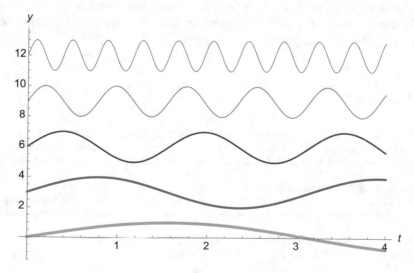

Fig. 5.27 Various waves obtained by doubling the period of the initial wave obtained using Zeleny's http://demonstrations.wolfram.com/PeriodAndFrequency/

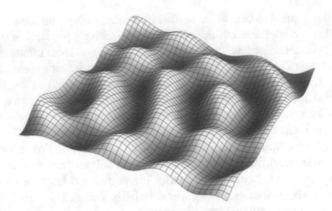

Fig. 5.28 A three-dimensional (the three axis omitted) wave generated by two sources roughly opposite to each other developed using J Nochella http://demonstrations.wolfram.com/3DWaves

sistent with the discussion throughout this book. We have focused on water-related examples because we think that most people have familiarity with it. This limitation can be changed to other physical hazardous media such as air and soil. A central theme is the representation and assessment of exceedances: these can be critical in the general context of establishing policy thresholds. The reader can conceptually change water flows with other specific hazardous events. In most cases, analogies should readily come to mind. Although we discuss and assess uncertainty through probabilities and distribution functions, we also exemplify a simpler alternative, uncertainty intervals, as a means to represent uncertainty when the available K&I is insufficient to perform probabilistic analyses. The simplicity of these calculations may be sufficient to decide if a more complete probabilistic analysis should be conducted and at what cost. Moreover, uncertainty intervals are often inherent to analyses based on risk matrices. The principle that the *best possible science* should support the assessment of the available choices may generally indicate that probabilistic analysis meets this principle.

References

N.M. Abbassi. http://demonstrations.wolfram.com/IllustratingTheUseOfDiscreteDistribution

Asquith W, JE Kiang, T A Cohn, *Application at-site peak-streamflow frequency analysis for very low annual exceedances probabilities*, US Geological Survey, Tech. Rep. 2017, https://doi.org/10.3133/sir20175038

J. Bolte's. http://demonstrations.wolfram.com/SampleVersusTheoreticalDistribution/

C. Boucher's. http://demonstrations.wolfram.com/PercentilesOfCertainProbabilityDistributions/

R.J. Brown's Wolfram demonstration, http://demonstrations.wolfram.com/ConnectingTheCDF AndThePDF/

FAA, Using Modern Computing Tools to Fit the Pearson Type III Distribution to Aviation Loads Data, DOT/FAA/AR-03/62, Final Report (Sept. 2003)

B. Hoskings, L-moments: Analysis and estimation of distribution using linear combinations of order statistics. J. R. Stat. Soc. Ser. B **52**, 105–124 (1990)

K.S. Lowder. http://demonstrations.wolfram.com/ClosedFormFullLifeCycleDistribution/

I. Mac Leod Mathematica Demonstration, http://demonstrations.wolfram.com/AnscombeQuartet/

I. McLeod. http://demonstrations.wolfram.com/StandardNormalDistributionAreas/

J. Mcloone. http://demonstrations.wolfram.com/DataSmoothing/

NAS, *Risk Analysis and Uncertainty in Flood Damage Reduction Studies* (NA Press, Washington, D.C., 2000)

E. Schultz Demonstration http://demonstrations.wolfram.com/DecisionsBasedOnPValuesAnd SignificanceLevels/ (Undated)

De Souza-Carvalho's Wolfram Demonstration "Correlation and Covariance of Random Discrete Signals" http://demonstrations.wolfram.com/CorrelationAndCovarianceOfRandomDiscrete Signals/

USACoE *Hydrological Frequency Analysis*, EM 1110-2-1415, 5 March 1993, Headquarters USACE, Washington, DC

S. Wolfram *Exponential Decay* Demonstrations http://demonstrations.wolfram.com/Exponential Decay/ (Undated)

E. Zeleny's. http://demonstrations.wolfram.com/PeriodAndFrequency/

A. Zheng. http://demonstrations.wolfram.com/MovingWaveAnalysis/

Chapter 6
Preferences, Choices, and Probabilistic Dominance: An Overview

6.1 Introduction

We discuss some of the critical concepts used in the evaluation of choices, such acting or not acting, given probable consequences. This material helps the analysis and evaluation of prospective choices and rationally informs decision-makers that seek either to minimize the adverse consequences or otherwise select between alternatives available to them. Understanding what catastrophic events can cause also implies assessing precautionary or preventive alternatives. The ranking of one choice over the other should be sound and unambiguous: it requires justification based on formal criteria. Although the dimensionality of a catastrophic incident can be very large, we simplify the discussion by assuming that it is finite and measurable. As we have discussed throughout the preceding chapters, a *dimension* implies directly or indirectly measurable data on a specific variable. For example, disposable per capita annual income (e.g., USD/person/year), either measured over time or as a cross section at a specific point in time, is a dimension of an *economic* subsystem that can be affected by a catastrophic incident. Physical events, from pyroclastic flow from a volcanic eruption incinerating thousands of people to ambient levels of concentrations [mass/unit volume of air] of $PM_{2.5}$ (emitted when producing energy and likely to cause chronic diseases), all involve dimensions. These are combined when formulating a causal model, one or more damage functions, depending on what is being analyzed. For example, a mathematical relationship quantifies the hypothesis that exposure follows response according to a specific function. Using two types of exposures, this hypothesis can be formulated to show that the risk factors act additively, using a linear model: *Response* $= a + b(Exposure_1) + c(Exposure_2)$. In this equation, a measures background (the effect of lack of exposure); b and c measure the change in response as the two exposures increases from zero. An alternative model is the multiplicative expression: *Response* $= c(Exposure_1) * (Exposure_2)$. The two risk factors act synergistically, their effect is greater than their sum. In these two contexts, dimensionality can be understood as follows. If the units of response are

© Springer Nature Switzerland AG 2020
P. F. Ricci, *Analysis of Catastrophes and Their Public Health Consequences*,
https://doi.org/10.1007/978-3-030-48066-0_6

#affected and those of exposure *mass of a toxicant/volume of air*, then *b* has units (*#affected*)/(*mass of the toxicant/volume of air*). Dependence and independence between variables in an exposure-response model have theoretical and empirical (e.g., statistical) foundations. The dimension of variables, for example, mass/volume of air, can be written in units of kilogram. [M], length in units of meter [L], time in units of second [T], and so on. In our example, concentrations have units of [M]/[L^3]. Dimensions characterize physical, chemical, or other quantities and may be fairly complicated, for example [person-year]. The practical limits of a study's K&I mean that some variables may be directly observable, others are indirectly measured (i.e., they are surrogate variables), and that the relationships between the variables are known with varying levels of certainty.[1] The dimensionality of inputs and outputs in a causal relationships implies having theoretically and empirically to consider the following:

- Completeness of representation of the alternatives, their full enumeration in terms of outcomes, state-of-nature, and measure of uncertainty such as probability
- Measurability of the dimensionality of inputs and outputs
- Decomposability of the system into its subsystems
- Existence and uniqueness of solutions
- Spatial and temporal plausibility (theoretical and empirical) regarding its history by accounting for structural changes in the subsystems that form it
- Model for or structure such as linearity or nonlinearity – based in part on criteria such as *Occam's Razor*, computability (e.g., computational time required to provide a result), and so on
- Revisions to the model's structure, conditioned on new information
- Revisions or retractions of assumptions and theoretical aspects and recalculations

The complexity of modeling catastrophic incidents can be defined by the interactions between its variables in the predictive, rather than descriptive, model that combines them. Moreover, data can be missing or unavailable; the temporal and spatial scales of events imply input variables that may be poorly understood. At the front-end, the choices concern finding the most appropriate model. At the decisional backend, the choices are the probable actions that best (in some formal sense) minimize the prospective magnitude of the adverse consequences. An example of a choice would be having to select between two exposure-response models. To begin with how to make choices, regardless of context, we assume that choices are rational

---

[1] In a statistical model such as R – a+bE, where R is response and E is exposure, the sets of data on *response* and *exposure* are called *vectors*; *coefficients* are often *scalar* quantities and can either be dimensionless or have units determined by the model. A *matrix* of dimensionality *j∗k* implies *k* dimensions and *j* observations on these dimensions. Different matrices will have different dimensions. It may be possible to reduce the number of vectors in each of these matrices by using statistical data reduction methods such as principal component analysis, PCA.

when they are axiomatic as developed by Ramsey, Kolmogorov, De Finetti, and others. The four axioms we think as critical are as follows:

1. *Completeness*: A P B, B P A, and A $\sim$ B. Here A and B are the two options, P means "is preferred to" and $\sim$ means "is indifferent to."
2. *Dominance*: If A P B in at least one dimension and is at least as desirable for all others, then A *dominates* B.
3. *Invariance*: The order of the presentation of the options does not change the conclusion. The introduction of an irrelevant alternative should not change the ranking.
4. *Transitivity*: If A P B than it cannot be that B P A (by the transitivity axiom: if A P B and B P C, then A P C).

Although much has been written about violations of these axioms, we propose that public decision-making must be rational because scarcity compels the maximization of the net collective benefits, when possible. Although paradoxes affect rational choices (such as Allais' and Ellsberg's), and because there are different rationalities, what rationality truly is has to be demonstrated. Different rules for making choices under time-of-the-essence constraints, such as *fast and frugal*, may be used (Gigerenzer and Goldstein 1996) instead of the methods that we discuss. Behaviorally, when dealing with the discounting of net monetary benefits calculated over time, individuals often prefer a certain sum now rather than a much larger some time later. Depending on the implications of public decisions associated with the choices being analyzed, a common criterion for choice is the maximization of the net expected monetized (and monetizable) benefits. This criterion may be replaced by the *max-min* rule which is not probabilistic and implies that some minimum level of well-being must be provided, namely, the maximum of the minima over a set of possible, mutually exclusive and nonoverlapping feasible choices. Although assessments are designed to inform decision-makers, they are separate from making policy decisions. Analysts are not public decision-makers: the latter are generally elected officials; the former are not. In this chapter, we are concerned with the former.

The component of the process leading to preferential or optimal choices may be summarized to five elements:

1. Formal choices, assumptions, and methods to represent causation should be parsimonious.
2. In public policy, value-based and analytical-based methods combine.
3. The assessment of equitable distribution of benefits and risks on those likely to be affected.
4. Reversals (e.g., between preferences) and other dilemmas or paradoxes should be explicitly addressed to increase stakeholders' confidence in the analyses and support policy decisions.
5. The aggregation of individual subsystems of the cause-effect relationships should not affect the behaviors expected from the causal system as a whole.

Figure 6.1 includes two methods for linking these five components while considering two different aspects of their uncertainty: probabilistic and fuzzy. This figure

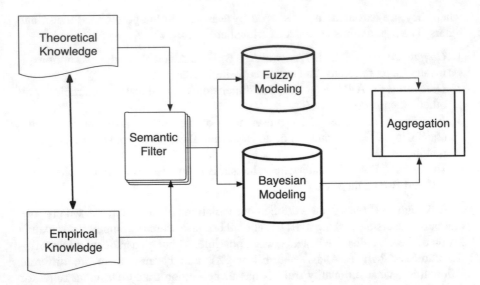

Fig. 6.1 Simplified flow of information and knowledge, a semantic filter to handle two different forms of uncertainty, their modeling, and aggregation of K&I to inform public choices

depicts how two different types of theoretical and empirical knowledge (fuzzy and Bayesian), first expressed in a natural language, can be formalized and processed analytically. We discuss in Chap. 7, *Integrating Heterogeneous and Uncertain Evidence*, these alternatives. Results from modeling are aggregated through systematic or other expert reviews and are judged by panels of experts through either ranking or voting methods developed in Chap. 8.

As Fig. 6.1 depicts, qualitative explanations are critical to decision makers. Boot and Thakor (2003) ask: To what extent *should the decisionmaker's intuition be allowed to supplant fact-based decision-making?* We think that the question is less critical in private than it is in public decision-making. Prospective public decisions – regardless of their formal basis – cannot be complete and certain. The same issue affects private decisions. However, private decisions generally involve spending privately held funds. Their owners may freely take chances of success without being concerned with the effect of failure (e.g., bankruptcy) on the public at large. This suggests that intuition can guide private decision-making. Public decision-makers, unlike private decision-makers, should have a common basis of information and knowledge that, in principle, is shared symmetrically by all parties (i.e., affected and unaffected). Intuition, within the context of deliberate and rationally justified ranking of public choices, cannot be dismissed but we suggest that it has a minimal role in public policy, relative to private decisions: rational analysis should guide public choices. Yet, the results may not be correct after the fact: the analyses may have been, for one reason or another, unrepresentative of the true problem. For prospective catastrophic incidents, different prior beliefs, theories, and likelihoods will give different views of the problem being addressed. For example, even assuming unbiased analysts, common background knowledge may not lead to the same beliefs because background knowledge can be understood differently by different experts. These may well use different distributions (*priors*) and different likelihoods (experimental results as empirical distributions) because of personal attitudes toward a particular scientific issue. It may also be the case that new evidence contradicts an accepted view and – even though it may be correct – it is ignored. Alternatively, new evidence may simply be hypothesis-generating rather than causal. Against this background, we develop the critical aspects of the analysis of probabilistic choices.

6.2   Making Probabilistic Choices: Lotteries, Time, and Dominance

Assessing prospective choices implies: (i) developing the choices; (ii) establishing the *states-of-nature* that interfere with them by not being under the control of the decision-maker; (iii) developing probabilities representing the uncertainty associated with the states-of-nature; and (iv) assessing the consequences (Clemen and Winkler (1999)). Although mathematically formalized, the basis for this assessment is behavioral and mechanistic. Each choice is deterministic. The set of choices should be complete, exhaustive, and nonoverlapping.

Kahneman and Tversky (2000) have introduced *decision weights* for probabilities to reflect the empirical findings that individuals overvalue small probabilities and undervalue large ones. There are other behavioral issues. For example, a sure gain is preferred to a lottery that has a larger expected value than the sure gain. This empirical finding contradicts expected utility theory and rationality: it violates the criterion that the maximization of the net expected benefits should drive rational individuals' choices. People may be more sensitive to difference in magnitude of gains or losses than to the magnitude themselves. Another issue is that most respondents opt for *pseudo-certain* outcomes. For instance, these two authors found that respondents having a choice of insuring against loss from fire and flood allocating *50%* of the premium between these two prefer to pay *100%* of premium to protect against one only. Under conditions of uncertainty characterized by probabilities, a probabilistic situation can be described by a simple *lottery* (L_i), in which the sum of the independent probabilities must be unity. The term lottery is understood as a probabilistic choice. A simple lottery consists of two mutually exclusive prospective outcomes (e.g., the alternatives *win* and *loss* are measured by money); each is weighted by the probability of either occurring or not. In the binary example (either a win or a loss), these two probable outcomes are mutually exclusive and fully encompass all that can occur for the purpose of this discussion. More generally, each of the choices is an outcome or consequence that can be positive, negative, or neutral. The criterion for choosing between alternative lotteries is to select the lottery with the largest expected net benefit, the minimum expected loss or the maximum expected benefit.[2] The issues that affect this *rational* analysis (Kahneman and Tversky 2000) are thoroughly discussed by these authors. We summarize five of them from Kahneman and Tversky (2000), Thaler (1981), Heukelom (2007), and Manski (2007).

1. *Sensitivity to Presentation of Alternatives* – Suppose having to make a decision represented by two probabilistic choices L_1 and L_2: $L_1 = 0.7*(1000$ prompt deaths) $+1/3*((0$ prompt deaths) $v.$ $L_2 = 700$ *deaths* (with probability $= 1$). Although the expected value of each lottery is the same (700 deaths), most of the individuals to whom this hypothetical was presented selected L_2 over L_1. If the alternative choices are stated (*framed*) as *lives saved* rather than *deaths averted*, and the lotteries are $L_3 = 333$ lives saved (with probability $=1$) $v.$ $L_4 = 0.33*(1000$ lives saved) $+ 0.67*(0$ lives saved), the majority of the respondents opted for L_3, rather than L_4. The second set of lotteries shows that respondents (who not at risk) are sensitive to the wording (*framing*) of the discussion, even though they should be unaffected because the expected value of each of the two choices is the same.

2. *Anchoring* – Preferences are influenced by the availability of an initial value because it can bias the response by those surveyed.

3. *Sequential Availability of Information* – Many decisions regarding the acceptability of a choice, and its perception, are sequential. Kahneman and Tversky (2000) illustrate how respondents may not be *rational* in their perception of a choice involving a sequence of prospects (P_i) by not maximizing the expected value of the prospects, when the consequences are positive; and conversely for negative consequences.

4. *Time as Discounting Factor* – For transactions based on money [\$], the traditional present discounted value rule[3] is PDV $[\$] = A[\$]/(1 + r)^t$ where A is the initial sum of money; r is the (discrete) discount (or interest) rate; and t is the number of periods used in discounting. The future discounted value, FDV $[\$] = A[\$]*(1 + r)^t$. The behavioral assumptions inherent to the PDV formula change are discussed in Thaler (1981).

Heukelom (2007) states that, relative to traditional discounting, "… individuals' intertemporal choices can be shown to be fundamentally inconsistent." More correctly to deal with the perception of future events, the *effective discount factor* (Heukelom 2007) consists of the sum of two components, the *"long-run discount factor d"* and the *"short-run discount factor βd,"* where $β < 1$. The traditional discount factor is explicitly decomposed in a long-run, exponential component, and a short-run, hyperbolic component (Camerer and Thaler (1995)). Individuals faced with *stochastic income* and a *borrowing constraint* anticipate their future inclination to hyperbolically discount (and to overconsume), and act against it. As Wilkinson and Klaes, (2008) note, newer discounting theories include subadditive discounting

---

[2] Probabilities are not cardinal in the sense that certainty (probability $= 1$) is not twice a 50% chance. The simplest *expectation* is $\mathrm{pr} * x_1 + (1 - \mathrm{pr}) * x_2$, the units of x_1 and x_2 is money; $\mathrm{pr} + (1 - \mathrm{pr}) = 1.0$.

[3] t is an index that enumerates the number of periods being discounted.

Table 6.1 Determinfrom dominance between alternatives, a_1 and a_2, given six states of nature (s_1 to s_6)

	s_1	s_2	s_3	s_4	s_5	s_6
a_1	12	15	40	40	40	40
a_2	12	15	30	30	40	40

for choices involving short (with higher discount rates) and long periods of time, a behavior that appears to occur not only in humans but to certain other species. Paradoxically investors tend to hold on stock that loose value, relative to their price of purchase, but sell stocks that increase in value, against the same benchmark (Barber et al. 2009).

5. *Dominance* – If a choice is at least as good as another and is better in one (or more) of their common aspects, then the former dominates all other choices considered. Inferior choices can be excluded. Table 6.1 depicts two alternatives, a_1 and a_2, six attributes, s_1 through to s_5, and 12 cells (a_i, s_j) containing the magnitude of the ordered outcomes (italicized). A choice is predicated on the premise that, in general but not always, more that *more is better than less*.

In this example, a_1 is everywhere preferable to a_2: a_2 can be eliminated because it is *ordinally dominated* by a_1 because a_1 yields larger outcomes than a_2, (a reward of x units or greater is at-least-as-likely under a_1 than under a_2), then a_1 dominates a_2. Suppose that we can functionally relate actions to values, $a = f(v)$, here v is a suitable measure of value to the stakeholders?[4] Consider a subset of actions, {A}, $a \in A$. We can assume some function that relates choices to values: $f(v) \in F$. Manski (2007) looks at *choice under ambiguity* as follows. A choice a_i is dominated by another choice a_j if $f(a_i) \leq f(a_j)$, for all $f(.) \in F$, and $f(a_i) < f(a_j)$ for some $f(.) \in F$. He adds two conditions: (1) $g(a_i) = g(a_j)$; and (2): $g'(a_i) > g'(a_j)$ and $g''(a_i) < g''(a_j)$. Under condition *1*, there is no ambiguity: a_i and a_j are equivalent (the decision-maker should be indifferent between them). But, under condition *2*, the choices are ambiguous: a_i and a_j cannot be ordered: no preference relationship (e.g., *more is preferable to less*) can be asserted. For example, suppose we consider a new action, a. Because of condition *2*, augmenting the choices from a and d to include e can either make things worse or better. However, we cannot tell which is going to be

---

[4]Following Manski (2007) ambiguity, ex ante of the event, occurs when the objective function (which formalizes what is to be minimized and the dimensionality of the objects for the minimization) is partially known. The objective of the analysis, and thus its dimensionality, is context dependent. The minimization of the adverse consequence ranges from deaths to property loss, from ecological damage to socioeconomic dislocations. For human-caused technological catastrophic incidents (e.g., Deepwater Horizon, British Petroleum drilling rig spill, fire, and explosion, Gulf of Mexico, 2010), the variables in the function is (or should be) known. This is also the case for technological disasters with extreme human consequences (e.g., Bhopal plant, methyl isocyanate toxic gas releases, Union Carbide, India, 1984). For extreme natural catastrophic incidents, this may not be the case (e.g., Puerto Rico effects of Hurricane Maria in 2017) due to the large dimensionality of the catastrophic event, the diffuse sources of the adverse consequences due the fragility of the areas affected, installations, and the type of individuals at risk.

which because of the ambiguity between $g'(.)$ and $g''(.)$. In the context of producing an output, if $g'(.)$ occurs there is an increase, but there is a decrease in output if $g''(.)$ were to occur. In practice, optimization is constrained by budgets and other factors and is generally modeled by a variety of methods in which an objective (stated as a function) is optimized (either maximized or minimized). Generally, for technological systems, the objective function, $f(.)$ (e.g., representing net economic profits) to be optimized may be linear or nonlinear. There is large number of deterministic optimizations methods, ranging from linear and nonlinear programming to dynamic programming. If the objective function to be maximized is not known explicitly, but is known that $f(.) \in F$, then the preferred choice is given by Manski (2007) who also discusses the limitation of the solution (in conditions of partial or complete ignorance). The minimization of probabilistic choices dictates that the option with the smallest expected number of adverse outcomes should be preferred.

6.2.1 *Probabilistic (Stochastic) Dominance*

We consider two simple lotteries, L_1 and L_2. If L_1 is more likely to result in an outcome at least x units greater than L_2, then L_1 dominates L_2, in the sense of first-order stochastic dominance, FSD, Fig. 6.2. These two lotteries imply two mass distribution functions in which F has probability pr and H has probability $1 - pr$. If $F \geq G$; $L_1 \geq L_2$, in which the symbol $\geq$ means "it-is-at-least-as-preferred-to-as..." and pr ($0 \leq pr \leq 1$). This is a formulation involving of *weak ordering*, $\geq$, *continuity*, and *independence* axioms. A lottery L_i has expected value $pr*F + (1 - pr)*H$; of course, only one of the two branches will eventually occur.

A more general issue is comparing two (cumulative) distributions, for example, two alternative and competing formulations in the context of dominance – whether

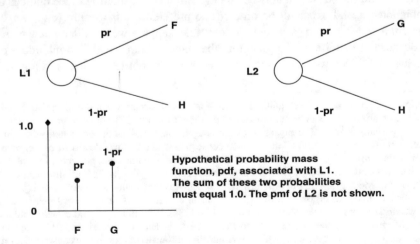

Hypothetical probability mass function, pdf, associated with L1. The sum of these two probabilities must equal 1.0. The pmf of L2 is not shown.

Fig. 6.2 Simple binary lotteries and their representation as mass probability distributions

one fully or partially dominates the other. That is, the distribution, *F, first order stochastically dominates, FSD,* the alternative distribution *G.* It can be proven that $V(F) \geq V(G)$ is equivalent to $F(x) \leq G(x)$, for all *x.* Thus (using some other conditions not provided for brevity's sake) *F FSD G* iff $V(F) \geq V(G)$, for any value of *x* ϵ *X.* First-order stochastic dominance is summarized in Fig. 6.3, using two continuous cumulative distributions. In this illustration, A_1 (represented by the cumulative distribution $L(X)$) dominates A_2 (represented by the cumulative distribution $G(X)$), for all $x \epsilon X$; namely $\int G(X)dX \geq \int L(X)dX$.

Second-order stochastic dominance, SSD, may be important to resolve issue where cumulative distributions cross, as depicted in Fig. 6.4. In this figure CDF(*X*) means cumulative distribution; the phrase applies to both $G(X)$ and $F(X)$. A_1 SSD A_2 if the cumulative area under A_1 exceeds the cumulative area under A_2. The random variable with cumulative distribution $G(X)$ second order stochastically dominates the random variable with distribution $L(X)$, provided that: $\int_0^t [L(X) - G(X)]dX \geq 0$.

Table 6.2 suggests an alternative representation of lotteries. In the situation described by the data in Table 6.2, all that is known is that the decision-maker is risk averse. She faces two mutually exclusive and fully exhaustive choices, a_1 and a_2 and six probabilistic states-of-nature, s_1 to s_6, over which she no control. That is, those six states may be physical (e.g., different levels of exposure, from low to high). The magnitude of the consequences is millions of dollars, for example, *1* in the cell with *1* means $1 * 10^6$; *2*, in the cell with 2, means $2 * 10^6$; and so on.

Calculating the expectations, of the two choices, E(.), yields: $E(a_1) = 0.16*1 + 0.16*1 + 0.16*4 + 0.16*4 + 0.16*4 + 0.16*4 = 2.88$ (millions of dollars), and $E(a_2) = 0.16*0 + 0.16*2 + 0.16*3 + 0.16*3 + 0.16*4 + 0.16*4 = 2.56$ (millions of dollars). If the decision-maker were risk neutral, she would select the option with the largest expected value, namely, a_1. However, there can be uncertainty about each of these two choices themselves. Thus, for instance, to maintain a state of neutrality between these choices, she can form another lottery, $L(2.88, 2.56)$, assigning two equal probabilities (0.5, 0.5) and calculate the expected outcomes of

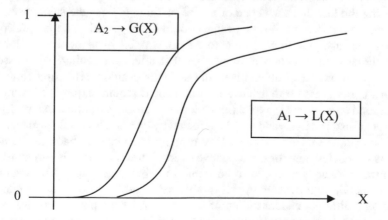

Fig. 6.3 First-order stochastic dominance between two cumulative distribution functions

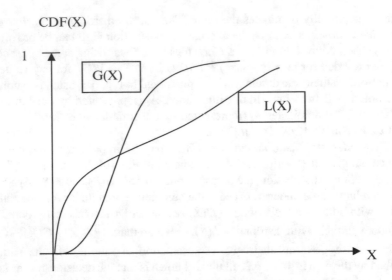

Fig. 6.4 Second-order stochastic dominance for two cumulative distribution functions

Table 6.2 Panel depicting two hypothetical alternative choices, a_1 and a_2, states of nature s_1 to s_6, consequences (integer numerical values), and probabilities (their sum = 1.0)

States of nature → Choices↓	s_1	s_2	s_3	s_4	s_5	s_6
a_1	1	1	4	4	4	4
a_2	0	2	3	3	4	4
Probability	0.16	0.16	0.16	0.16	0.16	0.16

The consequences are measured in millions of US dollars)

$L(.)$: $0.5(2.88) + 0.5(2.56) = \$2.72$ *(millions of dollars)*. The assignment of two equal probabilities to each lottery depends on the decision-maker's prior beliefs about these two choices. In this instance, she finds them equally probable, perhaps following the fact that each state of nature in Table 6.2 has the same probability. This example raises several issues that should be discussed as an integral part of the complete analysis. In general, before a catastrophic incident, there will be several alternatives and states-of-knowledge that affect the reality being assessed. Once a plausible set of alternatives and states is decided, retracting from them, perhaps under pressure to achieve results, should require extensive justification. The second issue is that the cells (a_i, s_j) contain values determined externally from the probabilistic analysis and may be affected by biases. These values are shown to be static; yet, the uncertainty that they represent can change because the assessment is conducted now for a prospective event. Third, probabilities are unlikely to be known with accuracy or even be stable over time, particularly for events that may happen in future generations. Another set of probabilities may be generated based on a different understanding of the structure of the physical reality as time progresses without the catastrophic event occurring. If *utilities* or monetary values

Table 6.3 Actions, states of nature, and utilities (integer number from 0 to some arbitrary upper numerical real number bound)		s_1	s_2
	A_1	1	4
	A_2	2	2
	A_3	4	1

of the consequences are known, but the probabilities are not, *dominance* can be used to identify and eliminate an inferior choice. Specifically, if the expected value of a lottery L, $[E(L)]$, is less than the expected value of a mixture of lotteries, L_1 and L_2, then that lottery can be eliminated from consideration. Using pr to represent probabilities and the expression $pr*E[U(L_1)] + (1 - pr)*E[U(L_2)]$, consider Table 6.3. The outcomes are measured by integer utilities, which we use to represent increasing levels of satisfaction. A utility may be thought of as a measure of value that does not exhibit diminishing marginal values (the more the value, the less the the satisfaction it produces) like money.

Let $pr(S_1) = p$ and $pr(S_2) = 1 - p$. The expected utility of $A_1 = p(1)(1)) + (1 - p)$ $(4) = 4\text{-}3p$; the expected utility of $A_2 = 2$; the expected utility of $A_3 = p(4) + (1 - p)$ $(1) = (p - 1)(3)$. If A_2 is preferred to A_1, the 2 must be greater $4\text{-}3p$: p must at least $2/3$. If A_2 is preferred to A_3, then A_1 must be preferred to A_2. Thus, an analyst should prefer either A_1 or A_3 to A_2. A_2 can be eliminated from consideration.

The *sure-thing principle* (Pearl 2016) is defined for an individual who will select between two acts, a_1 and a_2, such that if $a_1 > a_2$, given an external fact common to both, the preference relationship does not change regardless of whether that common factor occurs or not. This principle can be used to separate *acts* (prefer *acting a* to *acting b*) from *propositions* (prefer *finding a* from *finding b*): the sure-thing principle does not apply to the latter. Additionally, the choices made by the analyst do not affect a common, known fact. However, this assumption can be incorrect but can be remedied by letting *states* be *independent of acts* (Jeffrey, in Pearl 2016). Pearl's sure-thing principle states that if $pr(B|do(a_1)) = pr(B|do(a_2))$, and if an individual prefers a_1 to a_2 regardless of what the individual knows about some other fact B, then normatively $a_1 > a_2$.[5]

6.3 Conclusion

We base our discussion on the concept of lotteries, and the distributions that they represent, between alternatives because these objects are commonly understood and account for the triplet choices, consequences, and uncertainty through well-established methods. Public choices should be based on formal analyses to demonstrate the superiority of one choice over the others in the portfolio open to decision makers. Accounting for uncertainty, as discussed, is an important aspect of public

---

[5] The *do* operator (Pearl 2016, p. 239) implies epistemic causation. For example, in Bayes' theorem, it means changing from $pr(Y|X)$ to $pr(Y|do(X))$, based on observations on Y and X.

decision-making because budgets and resources are limited. There is competition for the allocation of funds to competing societal goals as well as for money in general between the public and private sectors. Although we limit our work to relatively few established results, and thus do not discuss more recent theoretical ways to select between uncertain alternatives, the material we develop is critical to understand more complex analyses and newer theoretical developments. We also address some of the less discussed aspects of ambiguity between choices, as can happen when their distributions overlap and dominance between them becomes complicated because the distributions cross or their uncertainty bounds overlap. We note that the methods we discuss also apply to private choices: the key difference is that the legal responsibility and the very nature of private objectives relative to public objectives.

References

B.M. Barber, Y.L. Lee, Y.J. Liu, T. Odean, Just how much do individual investors lose by trading? Rev. Financ. Stud. **22**, 609–632 (2009)

A.W.A. Boot, A.V. Thakor, The Economic Value of Flexibility When There is Disagreement, Amsterdam Business School, CEPR Discussion Paper No. 3709 (2003)

C. Camerer, R.H. Thaler, Anomalies: Ultimatums, dictators and manners. J. Econ. Perspect. **9**, 209–219 (1995)

R.T. Clemen, R.L. Winkler, Combining probability distributions from experts in risk analysis. Risk Anal. **19**, 187–203 (1999)

G. Gigerenzer, D.G. Goldstein, Reasoning the fast and frugal way: Models of bounded rationality. Psychol. Rev. **103**, 650–669 (1996)

F. Heukelom, Who are the behavioral economists and what do they say? Tinbergen Institute Discussion Paper No. 2007-020/1 (2007)

D. Kahneman, A. Tversky (eds.), *Choices Values and Frames* (Cambridge, UK, Cambridge University Press, 2000)

C.F. Manski, *Identification for Prediction and Decision* (Harvard University Press, Cambridge MA, 2007)

J. Pearl, The sure-thing principle, Tech. Rep. R-466, UCLA CS. Feb. (Los Angeles, CA, 2016)

R.H. Thaler, Some empirical evidence on dynamic inconsistency. Econ. Lett. **8**, 201–207 (1981)

N. Wilkinson, M. Klaes, *An Introduction to Behavioral Economics* (MacMillan, eBook, 2008)

Chapter 7
Heterogeneous and Uncertain Knowledge: Beyond Probabilities

7.1 Introduction

Science-policy analysis formally combines: (i) heterogeneous and uncertain evidence (often a mix of several established, conjectured, and conflicting mechanisms); (ii) alternative representations of uncertainty and causal modeling via probabilistic, fuzzy, and possibilistic methods; and (iii) expert group decision-making through ranking and voting. The US EPA's (2014, hereinafter *NexGen 2014*) discussion of current challenges regarding disease causation is an example of the integration of heterogeneous scientific information. The key challenge is due to the heterogeneity of uncertain causal factors and mechanisms. *NextGen 2014* deals with the micro level factors and mechanisms of chronic diseases that can be the delayed consequences of catastrophic incidents or even the catastrophe itself. *NexGen 2014* states, "… the key pathways in the network of interactions among genes, cells, tissues, and organs that is needed to conduct predictive toxicology; (3) further characterizing human variability and how genetic makeup, preexisting backgrounds of disease and exposure, and adaptive or compensatory processes combine to influence population risks; (4) accounting for variables in test systems that can influence observed associations between molecular perturbations and disease outcomes … and (6) characterizing, in the best way possible, the uncertainties and confidence in risk assessments informed by new data types."

The citation generalizes by changing biological terms with terms appropriate to other mechanisms and outcomes. For example, the various units of analysis (e.g., *genetic, variability of tests*) can be substituted by terms such as *hydraulic gradient molecular perturbation* with *another physical term* sets the stage for studying how a perturbation (e.g., a set of forces that cause a perturbation) can prospectively affect populations at risk. The second aspect of the generalization is summarized by the UK *Blackett Review of High Impact low Probability Risk* (2011). It states, "Where data are sparse or uninformative, recourse to expert judgment may be necessary in order to quantify low probability event likelihoods and to scale the intensities of

© Springer Nature Switzerland AG 2020
P. F. Ricci, *Analysis of Catastrophes and Their Public Health Consequences*,
https://doi.org/10.1007/978-3-030-48066-0_7

such events. Formalized procedures ... have been developed and are being applied increasingly to a widening range of medical, technical and scientific risk estimation challenges ... . Applying a structured approach offers an avenue for reasoning about uncertainties and the possibility of putting sensible numbers to otherwise intractable probability estimates. However, it needs to be recognized that the process of eliciting judgments from a group of knowledgeable subject matter experts is a non-trivial exercise - one of the main challenges for a problem owner, seeking inputs from an expert elicitation, is the selection of his or her expert panel."

Ascertaining *NextGen 2014 best way possible* can be demanding in conditions of uncertainty and vagueness. On the one hand, knowledge and data mining algorithms can be used when data is plentiful. These analyses justify empirical cause and effect from seemingly unrelated data bases; however, their results may lack the theoretical basis for asserting that the findings are causal. On the other hand, computational and theoretical aspects, when data consists of extreme values and thus may be limited by cost or sampling size, can result in scientific conjectures. This is particularly true when findings are extrapolated well beyond the relevant range of the experimentally available data. The third critical comment concerns the independence of experts selected to serve on an agency's panel or board. For example, in the case of regulatory agencies, the US EPA *Peer Review Handbook* (2015, p. *v*) states that that review *of influential highly influential scientific assessment,* defined by the OMB's *Peer Review Bulletin*: "... is conducted by qualified individuals (or organizations) who are independent of those who performed the work and who are collectively equivalent in technical expertise to those who performed the original work (i.e., peers). ... Peer review is an in-depth assessment of the assumptions, calculations, extrapolations, alternate interpretations, methodology, acceptance criteria and conclusions pertaining to the scientific or technical work product, and of the documentation that supports them."

Moreover (US EPA, *Peer Review Manual*, 2015), aside from not having conflicts of interest or ethical issues, ... *the independent peer reviewer should not be associated with the generation of the specific work product either directly ... or indirectly.* Yet, in the same that the *NexGen 2014* report was issued, the Chair of the Committee on Science, Space, Technology, Congressman Lamar Smith of Texas, wrote to the Administrator of the US EPA, Gina McCarthy (Smith, 2014) asserting that: *[a] mong the current ... Ozone Review panel, **16 of the 20** members are cited by the EPA ... the Agency cites the work of these panel members **more than 700 times** in these regulatory science documents they are being asked critically to asses* (emphasis in the original). Congressman Smith also stated that the Inspector General of the United States found that (since 2000) 70% of the 20 members of the Ozone Review panel *were principal or co-investigators for EPA grants totaling more than $120 million.* Although it may be impossible for an agency to avoid using experts involved in the research it supported, fairness and transparency imply divulging the technical details of individual and group judgment as they provide their aggregate assessments. For example, a learned society that often advises the US government, the US National Academies of Science, NAS, (2015), produces *consensus reports* through committees of experts convened to study issues of national importance. Several checks and balances are applied at every step of the process to protect the *independence*, *objectivity*, and *integrity* of the NAS' reports. These are produced by experts who are *non-stakeholders* and use *the best available evidence*. Independence is a necessary but insufficient condition for developing unbiased, transparent, and replicable assessments of what may be asserted to be *the best possible evidence*. We pose and attempt to answer two questions. These are as follows: (1) What should be done to maximize transparency and fairness when conjectures and conflicting theories may be advocated by diverse groups of scientific experts? (2) How can *the best scientific evidence* inform policy-science if it lacks formal definition? Data, modeling, individual, and group choices attempt to answer these two questions. Yet, particularly for our work, there is a critical difficulty. As *NexGen 2014* states, "… traditional methods are not adequately addressing complex … risk assessment issues such as co-exposures from many environmental stressors or the potential effects of chemicals on people who might be more sensitive or susceptible." To improve on this inadequacy, *NexGen* details an approach consisting of three tiers in which Tier 3 is the most extensive; the first two require lower experimental and analytical capital. The third tier deals with *major-scope* decision-making about *influential* and *highly influential* federal decisions, as defined by the Office of Management and Budget, OMB. Table 7.1, adapted from *NexGen 2014*, summarizes key strengths and weaknesses of Tier 3 assessments using the carcinogen B[a]P. Exposure to this chemical (and other agents) can produce occupational or environmental catastrophes, the effect of which depends on the latency of the disease or diseases in those exposed. The contents of Table 7.1 can be generalized to assessing prospective catastrophic incidents by changing the italicized phrases to one or more physical contexts.

Even well-established causation used in regulatory analyses involves different forms of uncertainty. For example, *NexGen 2014 consensus* model for B[a]P and the cancer it causes includes vague terms (underlined): "… conceptually describes

Table 7.1 Strengths and weaknesses, *tier 3 major scope assessment,* for the chemical carcinogen B[a]P (developed from *NexGen;* Table 3, p. 15, quotations in italics) NexGen; Table 3, p. 15 - US EPA, Office of Research and Development, Next Generations Risk Assessment: Recent advances in molecular biology, (2014)

Aspects of a *major scope decision*: advantages and disadvantages – B[a]P	
Strengths	Weaknesses
Increases understanding of mechanisms leading to dose or exposure assessment	Uses *traditional data to anchor molecular estimates of risk*. This implies that biomarkers of risk in a system have the same causal significance in another
Characterizes population heterogeneity and variability	
Aids understanding the effect of less well-known agents, if these have similar mechanisms of action	*Nonhuman data with non-concordant tissue response are challenging to extrapolate* (convert) *to humans*
Can include automated surveys of the literature which are faster, cheaper, and may yield more complete answers that traditional methods, including meta-analyses	Published peer-reviewed material may be inadequate for prospective assessments
	Many sources of variability can lead to false associations (i.e., false positives)

the events that <u>might occur</u> when B[a]P enters the cell. Briefly, B[a]P binds to AhR, leading to upregulation …, which <u>might lead</u> to additional B[a]P metabolism to epoxides, and increased oxidative stress. … The <u>core processes</u> are the induction of DNA adducts, mediation of … signaling, <u>alteration</u> of translesion synthesis…. Based on the <u>network interactions</u>, DNA adducts are <u>believed</u> to be formed … (citations omitted)."

Phrases such as *might occur*, *might lead*, *are believed*, and *raising questions* must be uniformly and consistently interpreted by individual researchers and by groups of experts. This may be difficult because these terms are qualitative, vague, and inconsistent with probability theory. Probability theory relies on fundamentally crisp numbers. The assertion that an argument has pr = 0.95 of being correct is a crisp because the probability that it is incorrect necessarily is pr = 0.05. Assumption and calculations involve crisply defined distribution functions, as we have discussed; these probabilities have precise meaning. For example, asserting that $X \sim N(0, 1)$ admits no other interpretation. Although we have relied on probabilities and probability distributions, there are quantitative alternatives to account for imprecise language. In this chapter, we discuss and exemplify some of the better-known alternative representations of imprecise uncertain reasoning. In particular, we discuss fuzzy sets and Dempster-Shafer *possibilities*.

7.2   Cause and Effect

Over 130 cancer types affect humans. In 2019, data from the U.S. National Cancer Institute, indicate that the number of lung and bronchial cancer cases approximates 228,000; the deaths are about 143,000. The number of melanoma cases is about 96,500 and the deaths about 7,200. Breast cancer is the most common: it caused 271,000 new cases and killed approximately 42,000. Of those deaths, 500 were males and 42,000 females: this is the most common cancer in the U.S. Of course, the numbers of all cancers' deaths is in the hundreds of thousand per year (in 2018, there were approximately 601,000 deaths due to all cancers in the United States and 1.90 million in the EU). By definition, those deaths are catastrophic consequences. Most of those cancers are multifactorial and involve biologically complex diseases. Part of the complexity is described by the US EPA *NexGen's* mechanistic network model for B[a]P. At the micro level of causal analysis, the AOP for liver carcinogenesis, is a Boolean network, BN, that is mechanistically: "[b]ased on the gene expression changes and activating DNA adduct formation… that can be used to predict the activation of cell cycle progression … . In a BN model, system dynamics are simulated by a series of connected nodes where each node represents a gene/protein and the connections between nodes (edges) represent some type of action/inhibition relationship. … ." This network (*NexGen*, p. 46): "… cycles through different overall system states, based on changes in the state of each node in relationship to the other nodes over time. To test a hypothesized outcome (e.g., that cell cycle progression and translesion synthesis will be sustained once initiated), the BN model was simplified to represent just the DNA adduct/cellular proliferation processes. … Of interest here is the occurrence of stable states or attractors, that is, cycles of states that recur and self-perpetuate."

NexGen: (1) mechanism of action is the complete sequence of biological events that must occur to produce an adverse effect; ... defined as "sequence of key events and processes, starting with interaction of an agent with a cell, proceeding through operational and anatomical changes, and resulting in an adverse health effect"; ... AOP describes a "sequential chain of causally linked events at different levels of biological organization that lead to an adverse health...."

An AOP network consists of interrelated AOPs that represent the combination of events and pathways that underlie disease or disorder. Critically, NextGen, (2014) states that it: ... *found that the phrase* <u>*weight of evidence*</u> *has become far too vague ... and is of little scientific use. The present committee found the phrase evidence integration to be more useful* ... (emphasis added).

Networks use mathematical operators such as union, intersection, and negation. This allows combining individual elements of a sub-processes into a single causal network. Additional information, such as the approximate importance of a node (its *centrality*), can be measured by its size (radius is proportional to centrality), using adjacency and centrality data (e.g., matrices). A deterministic *BN* has n nodes, k hedges between nodes, and a rule (truth table) for each node that specifies the state of the node at $t + 1$, given the states of the connecting nodes at time t.

Example   The deterministic *BN*, $V = \{x_1, ..., x_n\}$, $xi \in \{0, 1\}$, consists of a set of n nodes of physical entities that are assumed to take either state 0 or 1 (hence there are 2^n states for a network with n nodes and two outcomes, 0 or 1). Physical interactions and controls are assumed to be known and fixed, once a BN is developed. Each of its nodes x_i has k_i parent nodes at time t: these control their child node at time $t + 1$: $x_i(t + 1) = f_i(x_{i1}(t), ...)$, $\{i_1, ...ik_i\} \subseteq \{1, ..., n\}$, in which k is the connectivity of the i-th node and f_i is a controller (Xiao 2009). The function $f = (f_1, ..., f_n)$ denotes the Boolean network, $B(V, f)$; the network state at time t is $\mathbf{x}(t) = (x_1(t), ..., x_n(t))$, with f controlling the state transitions, e.g., $\mathbf{x}(t) \rightarrow \mathbf{x}(t + 1)$, (i.e., $\mathbf{x}(t + 1) = \mathbf{f}(\mathbf{x}(t))$). A probabilistic Boolean network extends these deterministic aspects through r Boolean networks, $B_1(V, f_1), ..., B_r(V, f_r)$, with individual network selection probabilities $c_1, ..., c_r$ summing to 1.00. At time t, the physical entities at the nodes are regulated by one of the BNs; at $t + 1$ there is a probability q to change network: a *BN* is randomly selected (from r *BNs*).

If the *BN* is based on a transition probability matrix, it determines the long-run state of the system through Markov chains. Perturbations representing mechanistic changes imply that each physical entity can randomly change state (Xiao and Dougherty 2007). Finally, numerical solutions and simulations determine the size and type of the *basins* of attraction for the system's trajectories. These *basins* represent the temporal evolution of the states of the dynamic system, determine its overall behavior, and identify the magnitude of a physical effect, relative to a solution yielding a smaller basin of attraction (Pal et al. 2005). Figure 7.1 depicts a BN in which the nodes can be either on or off during the period of time ($t = 100$ units) (http://demonstrations.wolfram.com/BooleanNetworks/, (no author given) and a simulation in which $N = 50$ and $K = 2$. Each node is randomly assigned an initial

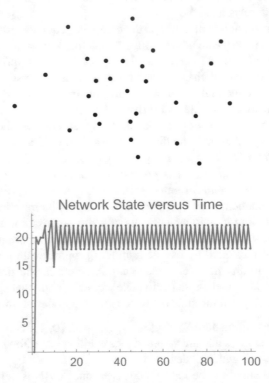

Fig. 7.1 Example of a Boolean network simulation in a specific state and its dynamics (state versus time) developed using http://demonstrations.wolfram.com/BooleanNetworks/, no author given

state and a rule table. Different values of N and K result in different network dynamics, which can be seen in the plot of network state (defined as Hamming's distance between the network's present state and initial state). Each node may represent a physically unique unit that can only be in or another state at some time $t + 1$. The dynamics of the network changes as the states (on, off) change according to the logic of its rules.

7.3   Integration and Fusion of Knowledge and Information

Public agencies have developed terms such as the *quality of the evidence, weight of the evidence*, and *mode of action* to deal with hazardous agents such as chemicals and ionizing radiation. In all cases concerning catastrophic incidents, the evidence of causation is quantitative, qualitative, and integrative (US EPA 2014; US NAS 2014). Modeling causally relates exposure inputs to relationships that range from fundamental physical laws to statistical associations with weak causality. For example, *NexGen 2014* (citations omitted) suggests two approaches to modeling complex (better understood as complicated) causation. One is the "bottom-up approach (that) focuses on molecular and cellular components, and seeks to understand how these components are networked, and how normal network function is altered following exposure to chemicals or stressors. …" The other is the "top-down approach (that) focuses on network interactions and disease indicators at the whole-body or population level, based often on human clinical and epidemiological data, and associations between disease states and environmental factors … Both the bottom-up and top-down approaches are informative and best used together to develop integrated and comprehensive knowledge." The bottom-up and top-down link causal modeling at the micro level to the macro level we have discussed by fusing particularized mechanisms and sub-processes to represent how the whole system responds to adverse exposures.

To achieve *integrated and comprehensive knowledge* through a systematic analysis, terms such as *aggregation, fusion, combination, composition, integration, synthesis,* and *propagation* permeate the scientific discourse. Like and unlike objects combine to form expressions that account for heterogeneous mechanisms and for the characteristics of the subjects to which those mechanisms apply to produce changes. In our work, an *aggregation* may be a simple mathematical operations that combine values from risk factors into one through single operations. *Fusion* is a higher level combination in which single and aggregated inputs such as distributions or fuzzy memberhip fucntions combine through a model, (e.g., Bayesian networks, fuzzy controllers). *Integration* combines both top-down and bottom-up approaches to yield the more complete causation from inputs and mechanisms to produce outputs (e.g., magnitudes of the adverse effects, predicted increase incidence), Fig. 7.2.

Figure 7.2 is a view of causation that leads to form a *summary of the confidence or reliability* and its *best possible characterizations* based on four sets of actions (from the highest (1) to the lowest (4)):

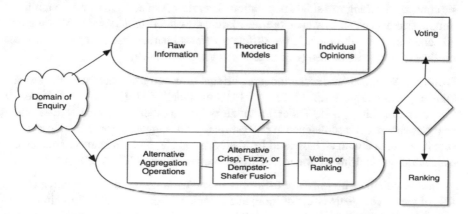

Fig. 7.2 Integration of micro and macro views and approaches into group judgment about an arbitrary reality (the domain of enquiry)

1. *Group Decision (agreement, consensus)* – Combines experts' opinions about existing evidence of causation for policy support (e.g., voting or ranking group choices).
2. *Causal Modeling* – Integrates from mechanistic processes to causal associations leading to predicted outcomes, conditioned on exposure, via a system approach.
3. *Knowledge Integration* – Combines intermediate sub-processes via differential or difference equations of various types (e.g., PDEs, stochastic calculus), networks (e.g., Boolean, Bayesian, fuzzy controllers), and statistical models (e.g., regressions, clustering algorithms).
4. *Aggregation of Raw Data* – Consists of compensatory or non-compensatory formulae.

These four characterizations depend on the evolution of information and knowledge depicted in Fig. 7.3 in which the context is modeling a disease specific process. Figure 7.3 depicts an overall process for assessing causation in the context of science-policy in which outcomes are levels of catastrophic health consequences.

7.3.1   Meta-Analysis

If the central issue is how to systematize existing evidence from the literature, there are practical alternatives, from which we select *meta-analysis* as an example. It consists of a portfolio of statistical methods used to assess empirical results from independent and essentially identical sampling studies generated by others. Meta-analysis captures the overall central tendency and variability of their results and combines them statistically. Meta-analysis has been applied to develop the distribution of the estimated coefficients of exposure-response models (the estimates reported by many authors of a very large number of regression equations) from several independent epidemiological studies and other studies. It can deal with fixed or random effect epidemiological models. A *fixed effect* model assumes that all studies conducted belong to a single population. Therefore, under this assumption, if the sample size of each study were extremely large, the effect studied would be equal in all samples. If, on the other hand, different effects characterize a population, the *random effect* model is used (Borenstein and Rothstein 1999).

Example We consider the carcinogenic effects of waterborne inorganic arsenic, a known *poison*. The NAS (2001, p. 14) concluded that (MCL is Maximum Concentration Limit, a US Federal regulatory law number): "[r]ecent studies and analyses enhance the confidence in risk estimates that suggest chronic arsenic exposure is associated with an increased incidence of bladder and lung cancer at arsenic concentrations in drinking water that are below the current MCL of 50 µg/L."

In this example, we only use estimates of the odds ratio, OR and relative risk, RR, of the carcinogenic effect of inorganic arsenic by Ferreccio and Chiou in 2000 and 2001 summarized in NAS (2001). We obtain their aggregate empirical

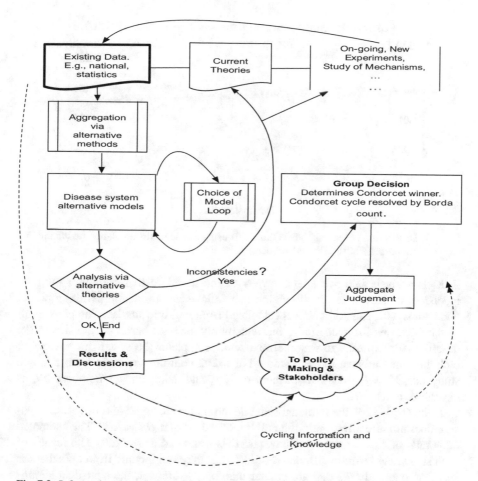

Fig. 7.3 Information, knowledge, and their cycling conditioned on evolving evidence

distribution, through a Monte Carlo simulation ($n = 10,000$), depicted in Fig. 7.4, in which three statistically insignificant studies are averaged with three others that are statistically significant implying compensatory averaging.

Figure 7.4 suggests that the average risk ratio (exposed to unexposed) is greater than 1, approximately 1.50, and therefore very low exposure empirically are shown to cause cancer. However, even though the numerical estimate shows increased cancer risk from exposure to waterborne arsenic, these ratios are insufficient to support an assertion of causation. Regarding whether inorganic arsenic exposure should be reduced from the then existing *50* μg/L standard, the US EPA did lower it to 10 μg/L (US EPA 2009). It relied on three panels of experts to assess the impact of this new regulation and the scientific basis to support changing the MCL from 50 to 10 μg/L. Those panels were formed by the National Academy of Sciences, the National Drinking Water Advisory Council, and the US EPA Advisory Board.

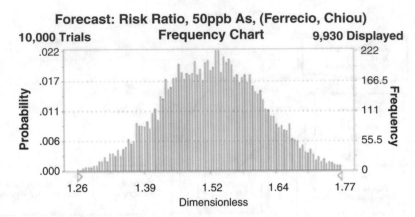

Fig. 7.4 Distribution of the risk ratios (Table 7.3) using Monte Carlo simulation (assuming that each risk ratio is probabilistically independent)

Meta-analysis is a systematic review of the literature that combines multiple independent studies. In the discussion that follows we are dealing with empirical odds ratios, OR, each of which is calculated from multiple studies. The plots of all of the ORs, one for each study, and uncertainty analysis based on the simulation described for exposure to waterborne arsenic, completes the systematic review by including the variance of each study. For a 2*2 contingency table from a single study the $OR = [(a*d)/(c*d)]$, where a, b, c, and d are integer numbers ≥ 0, as described in Table 7.2.

If the $OR = 1.00$, the numerator and denominator are equal and therefore exposure does not affect response: the null hypothesis is that $OR = 1.00$. The estimated odds ratio of each of the i-th study, $(OR)_i$, is estimated as $(ad/cb)_i$. The results of each study may be quite different. For instance, they can include from *ORs* that are less than one and *ORs* that are greater than one. Moreover, the statistical significance of those *ORs* can vary from study to study and can be calculated. Meta-analytic studies are observational because the meta-analysts cannot change the analyses done in the original studies: she takes those studies as she finds them. In other words, she has no control over the quality of the data and the statistical methods used by the original researchers, and so on. This analyst must: (i) identify and use the relevant studies; (ii) abstract the appropriate numerical information (e.g., coefficients of correlation, *ORs* or other) from them; and (iii) apply statistical methods to obtain aggregate results. Table 7.3 contains some of the steps for developing

Table 7.2 Contingency table for calculating statistical quantities such as the odds ratio, OR, relative risk, RR, likelihoods, sensitivity, specificity, and other statistical/epidemiological quantities

	Exposed	Not exposed
Response	a	b
No response	c	d

Table 7.3 Basic steps for conducting a meta-analysis; additional information in Borenstein and Rothstein (1999)

Steps	Description of step	Comment or question
State the research issue	Define the objectives of the meta-analysis	Is the result from the meta-analysis used for research or for policy?
Determine the sources of the original studies to be meta-analyzed	Literature searches may be limited to published and peer-reviewed papers to avoid the file drawer problem and the propensity for publishing studies that show positive results, rather than no statistically significant ones	These searches are often incomplete because of human limitations, databases limitations; errors of omission, researcher bias, and so on. Include government reports? Include trade publications?
Establish and justify criteria for inclusion and exclusion of studies	The included *and* the excluded studies must be reported and cited, with the criteria for exclusion clearly stated and attached to each excluded study	Sensitivity analysis can be used to study the effect of the exclusions on the meta-analytic results
Establish the statistical protocol for implementing the meta-analysis	Describe the tests and other statistical quantities to be used	Statistical methods and assumptions must be fully discussed and reported.
Assess the quality of the studies, weigh them, and conduct the meta-analysis	Independent analysts should verify the data base developed for meta-analysis and the analyses themselves	Discuss the results of the meta-analysis.

a meta-analysis, under the (strong) assumptions that the studies being used are independent and that all studies are available to the analysts.

The example that follows shows the input data to the meta-analysis and a summary of results using hypothetical data.

Example The example uses a data set available in the software program *Comprehensive Meta-Analysis* (Borenstein and Rothstein 1999). It consists of a literature review that includes the following results from 33 hypothetical studies of survival in which the analysis compares immediate treatment with streptokinase to prevent mortality following myocardial infarction to delayed treatment (treatment reduces blood clots). A summary of our meta-analysis of these 33 studies results is described in Table 7.4.

These three sets of results consist of three different tests. The Mantel-Haenszel test for fixed and random effect and Peto's test for fixed effects model only. The *OR* from the meta-analysis is less than *1.00*. The meta-analysis of those 33 studies shows that, taken together, immediate treatment is more favorable to the patient than delaying treatment. The width of the confidence interval may encompass the value of the null hypothesis: either *OR = 1.00*, or *ln(OR = 1) = 0*. The probability value (*p*-value) reflects this judgment. In this example, the *p*-value = 0.000 means that we are now dealing with probabilities lower than one in a thousand that the experimental

Table 7.4 Results from the meta-analysis of the survival data comparing immediate and delayed treatments

Meta-analysis methods	OR; 95% confidence interval (upper and lower confidence limits)			Null hypothesis, two-tailed test, 95% confidence interval	
Mantel-Haenszel	*Estimated OR*	*LCL*	*UCL*	*z-value*	*p-value*
1. Fixed effects	0.765	0.717	0.816	−8.158	0.000
2. Random effects	0.762	0.682	0.851	−4.839	0.000
3. Peto, Fixed effects	0.765	0.717	0.816	−8.176	0.000

result is due to chance. The computed *p*-value after data are collected and analyzed versus the decisional *p*-value adopted before doing the analysis. In other words, the null hypothesis can be either true or false.

T Garza's (2011) http://demonstrations.wolfram.com/ConfidenceIntervalsFor TheBinomialDistribution/ adds additional insights. He states that any confidence interval about a parameter of a probability distribution *must show two basic properties*. That is, it must encompass the value of the parameter with the decisional probability used and it should be as short as possible. In other words, suppose that the mean is *5*, if the standard deviation is *10*, the result is not particularly informative other than stating that the variability is much larger than the average. On the other hand, if the mean is *5* and the standard deviation is *0.05*, then the result is certainly informative because the variability of the results is much smaller than the average.

7.4   Knowledge and Information, Fusion

Combining different aspects of information and knowledge, knowledge and information, occurs through different mathematical operators, e.g., intersection (*Min*), averages (arithmetic, harmonic), and unions (*Max*):

1. *Conjunction* (e.g., union; *t*-norms for fuzzy sets; *Min*). Given partial scores, the aggregate score will be high iff the partial scores are high. The *Min* is (not strictly) monotone, symmetric, associative, and idempotent.
2. *Disjunction* (e.g., intersection; *t*-conorms for fuzzy sets; *Max*). Given partial scores, the aggregate score will be low iff partial scores are low. The *Max* is (not strictly) monotone, symmetric, associative, and idempotent.

These operators can be used to combine different objects, such as real numbers or labels; operators can be either crisp or fuzzy (Baas and Kwakernaak 1977; Chen and Hwang 1992). Aggregation functions can also represent attitudes: for instance, weights privilege some data over others, such as old data being weighted less than newer data in the context of a weighted average Xu (2005). Combining operations may be:

1. *Compensatory* (e.g., averaging operators). Low and high scores compensate each other.
2. *Non-compensatory* (e.g., symmetric sums). These functions are continuous, non-decreasing, and commutative.

Probabilities are aggregated using the methods from probability theory; probability distributions are practically combined through simulations such as Monte Carlo and Markov chain Monte Carlo methods. The Monte Carlo method is exemplified in the waterborne arsenic case. Aggregating functions and their characteristics are described in Table 7.5.

7.4.1 Fuzzy Modeling

A fuzzy measure, $g(.)$, is additive when $g(A \cup B) = g(A) + g(B)$ if $A \cap B = \emptyset$; super-additive when $g(A \cup B) \geq g(A) + g(B)$, and sub-additive when $g(A \cup B) \leq g(A) + g(B)$, under the condition $A \cap B = \emptyset$. A way replicably to formulate an assessment when it contains ambiguous or vague terms is essential when seeking an objective agreement that informs stakeholders. Fuzzy sets describe linguistic variables that contain vague concepts, for example, *much larger than* a crisp number: this statement is modeled by membership functions. Membership functions mapping from a crisp set to the interval [0, 1], given the context of the analysis. For example, a Gaussian membership narrowly concentrates high membership values, including 1.0, and two 0.9 values symmetric about 1.0: the phrase *a number close to X* = has semantic meaning. Notably, the membership function $\mu(x) = (1 + |x - a|)^{-1}$ has a cusp at a, $(\mu(x) = 1)$, and concentrates information differently from the Gaussian. *Hedges* (adverbs) can be modeled by: (1) dilation (taking the square root of the membership function), meaning *very*; or (2) concentration (squaring the membership function), for example, meaning *somewhat*. The combination of hedges and fuzzy sets yields much more depth and insight into semantic meaning: for instance, *very clear* is represented mathematically. Directly accounting for adjectives and adverbs gives a level of understanding that is consistent with the human descriptions via natural languages, unlike probabilities. An important aspect of membership functions is their shape. It can be determined through questionnaires, experiments, and expert judgments. Fuzzy aggregation of membership functions uses the basic operations of fuzzy logic. For two sets A and B (with complementation $\neg\mu_A(x) = 1 - \mu_A(x)$), (where x a crisp number), the union and intersection operations $(\cup, \cap)$ are, respectively:

- $A \cup B \rightarrow \mu(x)_{A \cup B} = \text{Max } \{\mu_A(x), \mu_B(x)\}$ for all $x \in X$.
- $A \cap B \rightarrow \mu(x)_{A \cap B} = \text{Min } \{\mu_A(x), \mu_B(x)\}$ for all $x \in X$.

For example, the truth, T, of the conjunction (A AND B) is the minimum of the truth values of A and B. Thus, T(A = exposed AND B = poor diet) = *Min* ((exposed) *AND* (poor diet)), implying a low health outcome. Truth, T, is not absolute: it is relative to the *universal set that contains both sets of A and B* values, is subjective, and

Table 7.5 Aggregation functions used with crisp and fuzzy numbers and implications (averages fall between results obtained via intersections and unions)

Properties of aggregation functions	Implications
Order statistics: neutral, continuous (Fodor and Rubens 1994)	*Neutral* means that an element in the aggregation should not influence the result. Can be equivalent to the Max, Min, Median, AM; the k-th order statistics is often used in policy analysis
Quasi-arithmetic mean: idempotent, neutral, continuous w.r.t each of its variables, f is strictly monotone. For example: $$M\left(x_1,\ldots,x_n\right) = f^{-1}\left[1/n\sum_{i=1}^{n} f\left(x_i\right)\right]$$	*Idempotent* implies compensatory average. *Compensation* implies that the aggregate can be < than the largest element or > smallest element. *Neutral* means that, given a set of inputs, a small change in one of them should not affect their aggregation. *Monotonicity* implies that if an element of the argument increases, the aggregate function must also increase. The means include the GM, HM, and AM
Associative averaging: idempotent, neutral, continuous, monotonic (Dubois and Prade 1988)	Averaging across inputs in different ways shoud not change the output average
Ordered weighted averages, OWAs: idempotent, compensatory, commutative, and monotone. $$\text{OWA}\left(x_1,\ldots,x_n\right) = \sum_{j=1}^{n} w_j\, x_{\sigma(j)}$$ in which σ is a permutation that orders the elements as $x_{\sigma(1)} \le x_{\sigma(2)}$; and the sum of the weight equals 1 (Yager (1996); Sadiq and Tesfemariam 2007)	Includes *Max, Min, k*-order statistics, median, AM, and the Choquet integral using specific weights
Choquet discrete (fuzzy) integral: idempotent, monotonic, compensatory, continuous. It is an averaging operator based on permutations: $$C_g = \sum_{i=1}^{N} x_{\sigma(i)}\left[g\left(A_{\sigma(i)}\right) - g\left(A_{\sigma(i+1)}\right)\right]$$ Where $\sigma_{(i)}$ is a permutation on N, with $A_{\sigma(n+1)} = 0$. Let S be a subset of the set T, s(S) measures the importance of S relative to T. If beliefs are super-additive, the implication is that the researcher is ambiguity averse and, as more credible information becomes available, aversion decreases although it will not equal zero even under certainty (Grabisch et al. 1998)	Aggregates ordinal values. Generalizes additive operators. Can contain Min and *Max* operators, and order statistics. Fuzzy integrals have two components: a weight associated with each factor, $g(\{x_i\})$, and a weight associated with each group of factors (the lattice structure of the groupings). Suppose that there are two risk factors: x_i, and x_j. The following may occur. If $g(\{x_i, x_j\}) = g(\{x_i\}) + g(\{x_j\})$, then these factors are independent. If $g(\{x_i, x_j\}) > g(\{x_i\}) + g(\{x_j\})$, then there is a positive (super-additive) interaction; reversing the inequality yields an antagonistic (sub-additive) interaction

[a]AM is arithmetic mean, GM is geometric mean, and HM is harmonic mean. Additional properties include stability and invariance with respect to change of scale

is context dependent. The degree of truth is measured in the interval [0, 1] via an appropriately selected membership function. Thus, IF proposition p is true to some degree, THEN proposition q is also true to some degree, but their joint degree of truth is not necessarily 1.00; the result is calculated via the *Max* and *Min* criteria.

For instance, assuming two sigmoid membership functions for $A(x)$ and $B(y)$, we can form the statement *IF x is A THEN y is B*, in which *x is A* has membership value 0.80 and *y is B* has membership 0.90. Using the *Min* rule, Min(0.80, 0.90) = 0.80, yields the context dependent degree of truth associated with $(A \text{ AND } B)$. Membership functions are descriptive: this is in sharp contrasts with distribution functions. On the other hand, probability distributions can theoretically depend on the nature of the random variable with which it is associated (e.g., rare events may imply the Poisson distribution or other extreme value distribution but not the normal distribution). Fuzzy integrals combine information by accounting for the power of the set to which the information belongs. Those integrals have two components: a weight associated with each factor, $g(\{x_i\})$, and a weight associated with each group of factors (e.g., a measure of the quality of the source of information). The Choquet integral is appropriate for ordinal values; this integral is idempotent, continuous, and monotonic. This integral is taken over discrete *capacities*; a *capacity* on a set of factors X, $X = \{x_i\}$ and a set function f: $2^X \rightarrow R^+$, $f(0) = 0$. This implies that the (membership) function $g(A) \leq b(B)$, A and $B \subseteq 2^X$; $g(A)$ can be used to determine the importance of A. In this case, Shapley's *importance* index (Murofushi and Soneda 1993) and the *interaction* index (Grabisch et al. 1998) give additional insights to fuzzy values. For example, if the importance index is >1, then the importance of the i-th factor is greater than average importance. If the interaction index is <0, then the i-th and j-th factors are antagonistic; if it is >0, these two factors are synergistic (Grabisch et al. 2009).

Example   Let sources of information be $\{X\} = \{x_1, \ldots, x_n\}$, experts provide $g(x_i)$, g: $2^X \rightarrow [0, 1]$, $g(\emptyset) = 0$ and $g(1) = 1$, monotonic. We may wish to find the fuzzy measures associated with the remaining combination of sources, namely, with the parts of the set $\{X\}$; Sugeno λ-measure provides the means to do so. For $\{x_1, x_2\}$, Sugeno's λ obtains from the equation:

$$\lambda + 1 = \prod_i^{\infty} \left(1 + \lambda g^i\right), \lambda > -1, \quad \text{in which } g^i = g\left(x_i\right). \tag{7.1}$$

For instance, let $g_\lambda(\{x_1, x_2\}) = g_\lambda(x_1) + g_\lambda(x_2) + \lambda g_\lambda(x_1)g_\lambda(x_2)$. With singleton fuzzy measures x_1, x_2, and x_3 provided by an expert to equal 0.2, 0.3, and 0.1 yields a quadratic equation with positive and negative roots. The negative root is discarded, the positive root is $\lambda = 3.109$: the worth of information is 0.687. The fuzzy measures for the other two combinations, $\{x_1, x_3\}$ and $\{x_2, x_3\}$ are 0.362 and 0.493, respectively. Hence, the worth of the information from these sources is greater than each of the singletons and less than that of the full set. Importantly, the entire worth of information is now known in its details, via a coherent and replicable method. The *level of support for the information* is a weight attached to the change in the permutated fuzzy measures of the worth of the information. Using the Choquet integral, the worth of the information is 0.262:

$$\int_{\text{choquet}} h \circ g = \sum_{i=1}^{3} h_{\pi i}\left(g\left(A_i\right) - g\left(A_{i-1}\right)\right)$$

(7.2)

$$= 1.0(0.1 - 0) + 0.3(0.362 - 0.1) + 0.1(1.0 - 0.362) = 0.262.$$

The term $g(A_i)$ is the worth of the information and $h(x_i)$ is the quality of the information. The sets are permutated (i.e., reordered from max to min):

- $A_i \underset{\pi}{=} \{x_{\pi 1}, \ldots, x_{\pi n}\}$;
- $h \to h\left(x_{\pi 1}\right) \geq h\left(x_{\pi 2}\right) \geq \ldots \geq h\left(x_{\pi n}\right)$;
- $g(A_1) = g(x_{\pi 1})$.

Both $h(.)$ and $g(.)$ are determined from experts, questionnaires, or other sources.

Fuzzy logic networks consist of rule (applied to hypothesis H) such as: IF *temperature is low*, THEN *incomplete combustion occurs*; IF *incomplete combustion occurs* AND *meat is on the burner* THEN *B[a]P forms*. That is, IF x is A, THEN y is B, with x and y defined over crisp sets. This reasoning can be extended by considering evidence E_i and outcome O_i, we may let:

IF E_1 THEN O_1.
IF E_2 THEN O_2.
...
IF E_n THEN O_n.

A computational expression for its result is: $\mu_H = \text{Max}[\text{Min}(\mu(E_1)), \ldots, \text{Min}(\mu(E_n))]$.

Inserting values for the minima and taking their maximum answers the question by yielding a degree of the overall truth of the available evidence, given the context of the analysis.

Consider natural language terms such as *inexpensive*, *close to work*, and *good suitability*. The rules are as follows: (i) IF *inexpensive* OR *close to work* THEN *good suitability*; (ii) IF *expensive* OR *far from work* THEN *suitability is low*; (iii) IF *average priced* OR *about 50 Km from work* THEN *suitability is regular*. A query would be: what is the *suitability of a house* that is *inexpensive* ($110*10^3$ \$) OR *close to to work* (57 Km)? This yield, under the OR operator: (0.45, 0.16); hence, Max = 0.45 that measures the degree of consistency between rules and antecedents. The reshaped aggregate (via the OR operator) consequent *suitability*, after defuzzification, yields the crisp number 4.85. Defuzzification (which returns crisp values from a fuzzy analysis) uses the centroid (center of gravity, COG) of the fuzzy output membership function. Inference is termed *aggregation* because it accumulates membership functions, via t-norms or co-norms applied to the membership functions on the RHS of these operators. LR Izquierdo and SS Izquierdo http://Demonstrations.wolfra.com/InferneceWithFuzzyIFTHENRules can be used to model these issues. Some fuzzy operations may be compensatory. The rules assume properties such as continuity (a small change in the value of an input variable results in a small change in the output variable), disambiguation (a unique value exists), and plausibility (there is a middle level of support). Fuzzy membership functions can include values that are more extreme than those contained in the data: a trapezoidal function accomplishes this.

7.4.2   Dempster-Shafer (DS) Measures and Their Combination

A DS measure is a probability mass assigned to the power of the set, $P(X)$. The key concept is *belief*, *Bel*, which represents a quantity measuring how much a given element of X belongs to the power of the set A: *Bel*: $P(X) \rightarrow [0, 1]$, (Wang and Klir 2009). The concept of *plausibility*, $Pl(.) = 1 - Bel(\neg A)$, is the dual of *Bel* and measures consistency with available evidence. The *basic* probability assignment, $m(A_i)$, measures the amount of evidence leading to believe that $X \in A$ (but not to what subset A_i may belong). It is similar to probability mass functions; probability measures involve similar axioms to D-S theory but require the stronger additivity axiom. Aggregation of these levels of support for the evidence from several sources can be obtained via Dempster's *rule of combination*. This rule is commutative and associative, but is not idempotent and continuous, which may create paradoxical results. For the two sets, B and C, their combination yields another set A (Yager 1987):

$$m_{1,2}(A) = \frac{\sum_{B \cap C = A} m_1(B) * m_2(C)}{1 - \sum_{B \cap C = \varnothing} m_1(B) * m_2(C)} \qquad (7.3)$$

Given a nested order of subsets, *Bel* and *Pl* are *consonant* over the body of evidence, E, and basic assignment m, if their focal elements are also nested (and thus free of dissonance). An element such that $m(A) > 0$, where $A \in P(A)$. For two sets A and B:

$Bel(A \cap B) = \min[Bel(A), Bel(B)], \forall A, B \in P(X)$.
$Pl(A \cup B) = \max[Pl(A), Pl(B)], \forall A, B \in P(X)$.

Example   Suppose two physicians independently examine a patient and agree that it can suffer from either meningitis (M), contusion (C), or brain tumor (T) (the *frame of discernment* is $\{M, C, T\}$. Assume that the diagnoses: $m_1(M) = 0.99$; $m_1(T) = 0.01$; $m_2(C) = 0.99$; $m_2(T) = 0.01$, thus yielding the combined results that $m(T) = 1$: the patient with certainty suffers from brain tumor. This paradox is due to Zadeh (1983). It arises because the diagnoses independently agree that the patient most likely does not suffer from a tumor but are in almost full contradiction for the other causes of the disease. Ali et al. (2012) modified the D-S combination rule obtaining $m(M) = 0.494745$, $m(C) = 0.49745$, and $m(T) = 0.0051$. This result is consistent with the intuition about the potential for T to occur, given the combined evidence from two independent sources of information.

DS's combination rule is commutative and associative, but not idempotent and continuous. Conflicting evidence can be discounted through Shafer averaging method: each *Bel* is weighted by the factor $(1 - \alpha; 0 \leq \alpha \leq 1)$ and then arithmetically quasi-averaged over all i beliefs. This neutral view, that all evidence should be accounted in an analysis and weighted for relevance, allows its overall assessment trough sensitivity analysis. A key criticism to DS theory is dealing with evidence characterized by severe conflicts (Dezert et al. 2012), known to yield paradoxical results under certain conditions (Zadeh 1983). These findings have resulted in several variants of DS's rule of combination, Table 7.6.

Finally, DS' *bpas* and the combination rules just listed apply to qualitative evidence, such as phrases *to a certain degree* and to combining them with other forms of evidence.

7.4.3   Bayesian Networks

Bayesian networks describe probabilistic conditioning between the variables of the network. The output of these networks is a joint probability, the product of conditional probabilities (or distribution functions as appropriate):

$$\text{pr}\left(Y_1, Y_2, \ldots, Y_k\right) = \prod_{i=1}^{k} \text{pr}\left(Y_i \,|\, \text{Parents}_i\right). \tag{7.4}$$

There must be *meaningful directionality* between the variables in the network. Testing *whether a proposed set of causal relationships is consistent with the available temporal-probabilistic information* can be obtained with Bayesian networks (Pearl 2009). An advantage of a Bayesian network arises from the *causal* meaning of the directed arcs. If S^1 represents a Bayesian network, S^2 another Bayesian network, and if $pr(S^1|D)$ is larger than $pr(S^2|D)$, then the change in probability makes the causal link stronger. However, assuming knowledge of the distribution of each random variable may not be consistent with the expressed distribution. An aspect of BN analysis that is useful in deciding between competing or alternative causal structures uses *stability* (meaning that there is an isomorphism between two competing structures) and *minimality* (meaning that the least complex of two structures, T_1 and T_2, given the same information and knowledge, is preferred). Minimality relates to the mathematical form and number of variables of the causal network. Stability relates to the probability of events in T_1 that make events in T_2 more probable; it refers to lack of extraneous, probabilistic conditional independences in a BN (Pearl 2009). This means that it is most improbable that the two competing structures overlap: a causal structure is minimal relative to a larger set of potential structure if no element of the minimal structure is preferred over those of any other structure in the class of structures considered (Pearl 2009). Hence, *minimality* and *stability* test the uniqueness of the network. Bayesian networks rely on directed acyclic graphs (DAG) to represent relationships between random variables. The term *acyclic* implies no feedbacks (Pearl 2009); a relative strict condition that can be relaxed at the cost of additional modeling complexity. The structure of the graph (e.g., a node, an arc, another node, and so on) is based on known or hypothesized relations that describe probabilistic conditioning and thus the dependence-independence structure among the variables of the network. This approach should

Table 7.6 Alternative Dempster-Shafer combination rules

Combination rule	Key expressions	Implications	Comments	References
Yager	$q(A) = \sum_{B \cap C = A} m(A)m(B)$; Conflict is measured by: $q(\emptyset) = \sum_{B \cap C = \emptyset} m(B)m(C) \geq 0$, it is assigned to the universal set X such that: $m^{Yager}(X) = q(X) + q(\emptyset)$	$B, C \in \wp(X)$; $q(.)$ is the ground probability mass assignment. $m^{Yager}(X)$ measures the level of ignorance.	Quasi-associative	Yager (1996). Sentz and Ferson and Ginzburg (1996) detail converting between Dempster and Yager rules
Zhang	$m(C) = k \sum_{A \cap B = C} \left(\frac{|C|}{|A||B|} \right) m(A)m(B)$ Where k is the renormalization factor such that $\sum m(.) = 1$. A relation between two set is defined by their cardinalities, $|.|$: $r(A,B) = \frac{|A \cap B|}{|A||B|} = \frac{|C|}{|A||B|}$	Accounts for couples of intersecting sets (e.g., $A \cap B = C$) via their cardinality $|.|$	Accounts for two frames of discernment through $r(.)$, which is not unique. The rule is commutative but not idempotent, continuous, or associative (Sentz and Ferson and Ginzburg (1996))	Zhang (1994), Sentz and Ferson and Ginzburg (1996)
Dubois and Prade	$m(C) = \sum_{A \cup B = C} m(A)m(B)$	Accounts for disjunctive consensus	Commutative and associative, but not idempotent	Dubois and Prade (1988)
Ferson, Root, and Kuhn	Given n bpas,[a] their convolutive averaging (c-averaging) is: $m_{1...n}(A) = \sum_{(A_1 + ... + A_n)/n = A} \prod_{i=1}^{n} m_i(A_i)$	A mixing average that yield results consistent with the arithmetic average	Commutative and quasi-associative, but neither associative nor idempotent	Ferson et al. (1999)

[a]*bpa is a basic probability assignment*

represent the disease of concern, the probabilities that characterize uncertainty, their conditioning, and inference. However, experts who might provide the probabilities or probability distributions needed to build a Bayesian network may:

- Not be willing to provide them.
- Have difficulties in assigning probabilities when a new biological process or conflicting epidemiological results.
- Not have some of the necessary or sufficient information or knowledge available at the time of the analysis (but may become available later).
- Even when willing to provide probabilities, ignore De Finetti *dogma of precision* (Walley and Fine 1982): *for each event there is some betting rate which is regarded as fair: one would bet on either side (for or against the event) at that rate.*

7.5   Individual Expert Opinions and Group Agreement

Agreements on a multifactorial diseases' causation, such as those generated by committees of learned societies or scientific advisory boards, rely on the consensus of a group of duly vetted individual experts' opinions. That consensus is often understood as unanimous *accord*. However, particularly at very low doses or exposures leading to multifactorial diseases such as cancer, the available evidence, its weight, and theoretical understanding cannot be expected to lead to unanimous agreement (other than someone will agree to disagree). *Consensus* becomes improbable when there is no unanimity on the assumptions and mechanisms, empirical information has gaps, modeling is complex, and the results are relatively small against a much larger background. Fortunately, less stringent agreements can be formalized when the causal model adopted for public policy cannot be claimed to be unanimously accepted due to the state of the science and the potential for litigation. It seems obvious that searching for causation at the scientific level must be matched by the same level of rigor that generates the consensus choice. A limited search of governments' websites does not disclose the details needed to understand how consensus is reached methodologically in scientific deliberations that inform public policy. Consensus about the final results of the examination of the literature and supporting testimony is made available, including abstentions or dissents. However, our review does not find information on the following: (i) the voting criteria used, (ii) how votes are weighted, (iii) the methods used in aggregating multiple expert opinions, and (iv) the loci of divergences between individual expert's opinions regarding each expert's assumptions, models, and results, Fig. 7.5. We describe how to reach a formal aggregation of opinions over heterogeneous objects, from the certain to the conjectural, and their combination. An axiomatic, rank-ordering preference-based voting method, which leads to defensible and reproducible agreements, suggests that ordering by linear preferences is appropriate for aggregating scientific judgments and reaches a demonstrably defensible agreement, if it exists.

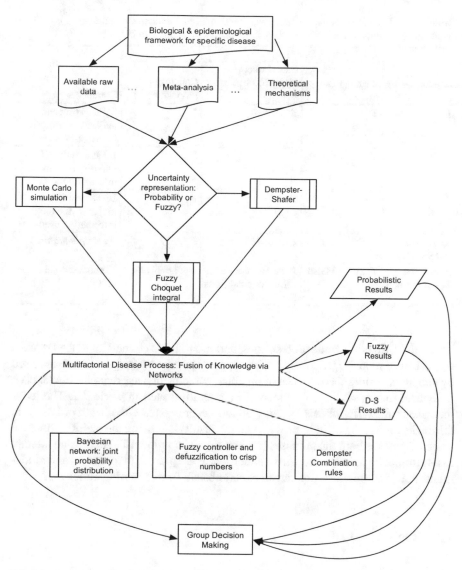

Fig. 7.5 Integration of measures and methods discussed in this chapter, informing group decision-making, while accounting for different forms of uncertainty

Table 7.7 depicts the elements of an instance where three experts are assessing the evidence before them. These experts are (i) presumed to be independent, (ii) have access to the entire set of alternatives by full access to the internet and the scientific literature, and (iii) vote on all of the elements of a common causal argument. What is assessed consists of: antecedents (assumptions), logical connections (model), and consequents (results). We use Boolean (true (T), or False (F)) voting

Table 7.7 Individual opinions on premises, causal rule, and conclusion can lead to conflicting aggregate judgments on each of these three components of a complete judgment

| Expert number | Judgments, J_i On: | | | |
	Assumptions: p q, are:	Causal rule: If (p AND q) Then r is:	Conclusion: proposition r is:	Comments
1	T	T	T	By assertion and majority rule on information used
2	T	F	F	By assertion and majority rule on information used
3	F	T	F	By assertion and majority rule on information used
Aggregate judgment	*T: (2Ts, 1F)*	*T (2Ts, 1F)*	*F (3Fs)*	*By simple majority rule*
Notes and key reference	*T*: true; *F*: false	Joint necessary and sufficient conditions for *r*	Not applicable	Kornhauser and Sager (2004)

alternatives, do not include probabilities, and use a single connections (AND) and a single causal model: *If … Then …* . The aggregate experts' judgments are assessed by simple majority rule over each of the three components: aggregation is over each individual judgment, J_i, via a function $F(.)$, as illustrated in Table 7.7. The judgments are summaries of each expert's opinion expressed as either truth or falsity. In this table, the aggregate judgment $F(.)$ on the results is F using the majority rule; the same rule makes both the premises (assumptions) and model T. There is no attempt to use formal logic in the cell where either T or F are placed; there are no other judgments (e.g., indifference) nor can an expert abstain from asserting either T or F.

7.6   Conclusion

Mathematical aggregation and fusion methods yield a synthesis of event-specific information and knowledge, accounting for uncertainty. Simplistically, aggregation applies to singletons through alternative averaging operators; aggregation through the Choquet integral extends these characteristics by accounting for the entirety of the power of their sets and two levels of judgment about the importance of its integrand. Fusion combines multiple aggregated elements and connects them through probabilistic or other causal networks to obtain knowledge that is directly applicable to science-policy. Importantly, natural language expressions regarding scientific causes, effects, and causation use adjectives and adverbs that often are fuzzy or vague (Chen and Chen 2005). In the instance of language fuzziness, fuzzy sets and their logic can be integrated through *fuzzy controllers*. We find that the two preferable knowledge fusion methods are (i) Boolean (deterministic and probabilistic) and Bayesian networks, and (ii) fuzzy logic controllers. The first two yield formal descriptions of linkages at different levels of granularity. For example, in the instance of system biology, the networks may be extended to include epidemiologic (human health) information to yield an integrated causal model. Regarding the joint effect of probabilistic variables, their joint distribution can be determined using Bayesian network. Fuzzy logic controllers yield *defuzzified* values (depending on the computational method used to obtain such defuzzification) based on inputs that account for linguistic modifiers, adverbs and adjectives, for those factors used in building the controller.

Consensus, or other form of aggregate judgment such as simple majority, informs administrators, ministers, and other decision-makers via expert panels convened by the government or learned societies. The four elements of this aspect of science-policy we summarize in this chapter are as follows:

1. *Group Decision (agreement, consensus, majority, or other rule)* – Combination of experts' opinions about existing evidence of causation for policy support to inform decision-making.
2. *Causal Modeling (networked sub-processes with certain or uncertain components)* – Integrates from mechanistic processes to causal associations leading to a disease outcome, conditioned on exposure, for instance via a system biology approach, from 3 and 4 below.
3. *Knowledge Integration* – Combination of intermediates sub-processes of a disease via differential or difference equations of various type (e.g., PDEs, stochastic calculus), networks (e.g., Boolean, Bayesian, fuzzy logic controllers), and statistical models (e.g., regressions, clustering algorithms).
4. *Aggregation of Raw Data* – Consisting of compensatory or non-compensatory formulae (e.g., averages, quasi-averages).

Acknowledgments This chapter was in part supported by a research contract funded by PMI, Neuchâtel, Switzerland.

References

T. Ali, P. Dutta, H. Boruah, A new combination rule for conflict problem of Dempster-Shafer evidence theory. Int. J. Energy, Info, and Comm. **3**, 35–40 (2012). www.cec.lu.se/upload/cec/BECC/BECCworkshop/BayesianMctaAnalysis.pdf

S. Baas, H. Kwakernaak, Rating and ranking of multiple-aspect alternatives using fuzzy data. Automatica **13**, 47–58 (1977)

M. Borenstein, H. Rothstein, *Comprehensive Meta-Analysis* (Englewood, 1999)

S.-J. Chen, S.-M. Chen, Aggregating fuzzy opinions in heterogeneous group decision environment. Cybernet. and Syst., An Int. J. **36**, 309–338 (2005)

S. Chen, C. Hwang, *Fuzzy Multi-attribute Decision-Making* (Springer, Heidelberg, 1992)

J. Dezert, P Wang, A. Tchamova, On the validity of Dempster-Shafer Theory, in *Proceedings of the International Conference*, Singapore, 2012

D. Dubois and H. Prade, *Possibility Theory: An approach to computerized processing of uncertainty*, Plenum Press, NY (1988)

D. Dubois, J.-L. Marichal, H. Prade, M. Roubens, R. Sabbadin, The use of the discrete Sugeno integral in decision-making: a survey. Int. J. Uncertainty Fuzziness Knowledge Based Syst. **9**(5), 539–561 (2001)

S Ferson and LR Ginzburg, Different methods are needed to propagate ignorance and uncertainty, Reliability and Engineering System Safety, 54:122–144 (1996)

S. Ferson, W. Root, R. Kuhn, *RAMAS Risk Calc: Risk Assessment with Uncertain Numbers* (EPRI, Palo Alto CA, 1999)

J. Fodor and M. Rubens, Fuzzy Preference Modeling and Multi-Criteria Decision Support, Klewer, (1994)

M. Grabisch, S.A. Orlovski, R.R. Yager, Fuzzy aggregation of numerical preferences, in *Fuzzy Sets in Decision Analysis*, ed. by R. Slowinski, (Wiley, New York, 1998)

M. Grabisch, J.-L. Marichal, R. Mesiar, E. Pap, *Aggregation Functions* (Cambridge University Press, Cambridge, UK, 2009)

G.J. Klir, B. Yan, *Fuzzy Sets and Fuzzy Logic* (Prentice Hall, Upper Saddle River, 1995)

T. Murofushi, S. Soneda, Techniques for reading fuzzy measures (III):interaction index, in *9th Fuzzy Systems Symposium*, pp. 693–696, Sapporo, Japan (1993)

NAS, *Arsenic in the Drinking Water, 2001 Update* (National Academies Press, 2001)

NAS, Division of Earth and Life Studies, *About Our Expert Consensus Reports*, 500 5th St., NW, Washington, DC 20001, (2015)

NAS, Advances in causal understanding for human health risk-based decision-making: proc. f a workshop in brief, NAS Press Wash. DC (2018)

National Academies of Science/Engineering/Medicine, http://dels.nas.edu/global/Consensus-Report, accessed 2-1-2016

NRC (National Research Council), *Review of EPA's Integrated Risk Information System (IRIS) Process* (The National Academies Press, Washington, D.C., 2014). Retrieved from http://www.nap.edu/catalog.php?record_id=18764.computations, *and system biology*, EPA/600/R-14/004 Washington, DC (2014)

R. Pal, I. Ivanov, A. Datta, M.L. Bittner, E.R. Dougherty, Generating Boolean networks with a prescribed attractor structure. Bioinformatics. **21**, 4021–4025 (2005). [PubMed]

R. Sadiq, S. Tesfemariam, *Probability Density Functions Based Weight for Ordered Weighted Averaging (OWA) Operators: An Example of Water Quality Indices,* NRCC-48680, (Institute for research and Construction, National Research Council Canada, Ottawa, 2007)

US EPA *Drinking Water Standards and Health Advisories Table*, Region IX, San Francisco, (November 2009)

US EPA, *Peer Review Handbook*, 4th Ed., Oct. 2015, EPA/100/B-15/001, Science and Technology Policy Council, US EPA, Washington, DC, 20460

US EPA, Office of Research and Development, *Next Generations Risk Assessment: Recent Advances in Molecular Biology,* (US Environmental Protection Agency, Wash. DC, 2014)

P. Walley, T.L. Fine, Towards a frequentistic theory of upper and lower probability. *Ann. Stat.* **10**, 741–761 (1982)

Z. Wang, G.J. Klir, *Generalized Measure Theory* (Springer, New York, 2009)

Y. Xiao, A tutorial on analysis and simulation of Boolean gene regulatory network models. Curr. Genomics **10**, 511–525 (2009). https://doi.org/10.2174/2F138920209789208237

Y. Xiao, E.R. Dougherty, The impact of function perturbations in Boolean networks. Bioinformatics. **23**, 1265–1273 (2007). [PubMed]

Z. Xu, An overview of methods for determining OWA weights. Int. J. Intell. Syst. **17**, 569–575 (2005)

R.R. Yager, On Ordered Weighted Averaging Aggregation Operators in Multicriteria Decision-Making, IEEE Trans. Syst., Man, Cybernetics, 18:183–190 (1988)

R.R. Yager, Families of OWA operators. Fuzzy Set. Syst. **59**, 125–148 (1993)

R.R. Yager, Quantifier guided aggregation using OWA operators. Int. J. Intell. Syst. **11**(49–73) (1996)

R.R. Yager, Constraint Satisfaction Using Soft Quantifiers, Unpublished manuscript available from the Author (yager@panix.com)

L.A. Zadeh, A computational approach to fuzzy quantifiers in natural languages. Comput. Math. Appl. **9**, 149–184 (1983)

Chapter 8
Aggregating Judgments to Inform Precautionary Decision-Making

8.1 Introduction

Limited knowledge and information, *K&I*, affect science-policy analyses of prospective catastrophic events. Those analyses combine expert judgments with empirical and theoretical evidence of cause and effect. If, however, evidence that is generally agreed to be critical to decision-making (e.g., epidemiological evidence) cannot be available, substituted evidence must be sufficiently probative to justify a regulatory decision. An important factor that limits the completeness of K&I is the length of period of time that lags between exposure to a hazardous condition for example, time to failure due to corrosion for structures, *incubation* and *latency* for chronic diseases. The scientific evidence used to formalize causation in regulatory law is probabilistic; it is also heterogeneous and based on experts' judgments. It provides the following: (i) the logical structure of a multifactorial, *n*-dimensional cause and effect process; (ii) populates it with factual evidence; (iii) develops cause-effect as probabilistic exposure-response models; and (iv) culminates in a review by a select group of experts that votes or ranks the strength of the overall findings to inform public policy. The policy-science question, given a specific context is: What amount of judgments should be sufficient to inform decisions? Here we focus on how expert judgments can be aggregated to produce this sufficiency, once expressions of causation become available from scientific theoretical and empirical modeling, the literature and from other sources of evidence, including seemingly unreliable sources. All of these elements may seem a tessellation. What is needed to make it a casual mosaic? The answer should include incomplete opinions, some of which may be equivalent to *hearsay*. Although seemingly unreliable, that evidence may be all that is available in the instance of unusual events or, otherwise, it may complement more substantive evidence. Unlike the more complete K&I discussed in our preceding chapters, this evidence is soft and may even be incorrect by design (as could be the case when one or more parties play strategic games and misinform each other). Nonetheless, in the aggregate, it should complement the understanding of risk factors, causation, and the magnitude and severity of the consequences (Wong-Parod et al. 2012; Dietz 2013).

8.2   Aggregating Opinions for Decision-Making

Expert opinions, transparency of process, and replicability are elements of aggregating experts to inform decision-making. This function is inherent to reviewing agencies' decisions. For example, the Congressional Review Act, CRA (5 U.S.C. Sect. 801–808), extends Congressional oversight of agency's *rules* by going beyond the scope of the Administrative Procedure Act, APA, and the scrutiny of rulemaking to include agency's guidance (e.g., *guidelines*) and other less formal documents. This statute requires that some federal agencies report to Congress, which can overturn regulatory actions through a joint resolution of disapproval (through established procedures between Congress and the White House (WH). The WH proposed an EO (in 2018) specifically aimed at the EPA, which states that this "*... proposed regulation is designed to increase transparency of the assumptions underlying dose response models. ... The use of default models, without consideration of alternatives or model uncertainty, can obscure the scientific justification for EPA actions.*" Yet, despite its specificity to the US EPA, the connection to other federal agencies is apparent because "*[a]lthough not directly applicable to other agencies of the Federal government, it may be of particular interest to entities that conduct research ....*"

Science-policy depends on expert scientific opinions. Expert individuals are the *filter* between scientific K&I and public decision-makers in the member states of the EU, the EU itself, and the USA at federal and state levels. The aggregation of those opinions has at least three functions: (i) assess the available heterogeneous and uncertain evidence, (ii) assess how the evidence is combined into a causal argument, and (iii) assist agencies to develop causal justifications for their scientific decisions. These functions are themselves interpreted and quantified by other special expert panels who assess the total evidence and decide which is the best available evidence to inform policy decisions. This later stage of expert aggregation implies determining the *correctness* of the available K&I (Straus et al. 2009; Hsiang and Meng 2014). We take correctness as the necessary quality of any aggregation of scientific evidence interpreted by scientific experts through (axiomatically sound) voting and ranking rules on the totality of that evidence with individual opinions (e.g., stated as *True*, *False*, or *Undecided*) on key elements of quantitative causal expressions. Panel voting or ranking yields a (relative) preferred choice, if it exists, which is presented to stakeholders in support of their decision-making (Von Winterfeldt 2013; Morgan 2014).

8.3 Choices by Voting or Ranking

Scientific evidence may be thought of as an *n*-dimensional tessellation whose boundaries are established from available evidence (e.g., toxic, genetic, and carcinogenic effects at low exposures of one or chemical, ionizing, or other substances). This tessellation is not evidence of the cause and effect that may be of interest. A mosaic of that evidence should however yield the actual, single substance basis for causation at very low exposures. Expert judgments construct the mosaic. Here, we focus on two formal aspects of integrating experts' judgments: voting and ordered preferential choices obtained through ranking. For both voting and ranking, we adopt an axiomatic approach, because axioms support demonstrating that causal assertions are (at least internally) correct. We take the *single decision-maker* as the final decision-making authority who is informed by the results from voting and ranking on the following: (i) assumptions, (ii) models, and (iii) conclusions. The simplifying assumption is that a single public decision is accountable and responsible to society in general, and the collective (those at risk) in particular. Most, if not all, regulatory science-policy contexts rely on integrating judgments, use direct and indirect evidence (i.e., the probative evidence just discussed) for their analyses, and select aspects of each from the following sets of K&I:

$$
\begin{aligned}
&\{\text{literature adequacy : e.g., summarizing systematic and} \\
&\text{automated findings of key characteristics of causal} \\
&\text{system}\} \rightarrow \{\text{adequacy of the research portfolio used} \\
&\text{by the public agency}\} \rightarrow \{\text{assumptions}\} \rightarrow \{\text{modeling}\} \rightarrow \{\text{results}\}
\end{aligned} \tag{8.1}
$$

Lay stakeholders may be involved in commenting on the results from the regulatory process. Although often neither scientists nor experts, they are heard during the *notice and comment* regulatory processes. The literature and ad hoc scientific studies and their results are interpreted by scientists who populate these sets by forming relationships between them (Morgan 2014). Subjective beliefs and biases that may affect expert judgments should also be formally characterized. Instructions of the experts regarding probabilistic or other analyses precede the elicitations of opinions before voting and ranking. The integration that satisfies scientific and policy requirements should include the following: (i) sound scientific assessments; (ii) formal rules, such as majority voting, on specific scientific propositions in the scientific assessments. Regarding the rules, collective objectives include procedural and objective fairness. For instance, rules may be:

1. *Proposition-wise voting* – ranging from plurality to absolute majority.
2. *Proposition-wise quota* – threshold based, may include a *distance* metric between the conclusion and the centroid of clusters of individual experts' choices.
3. *Conclusion-based voting* – ranging from plurality to absolute majority.

Plurality voting means that for any choice x and y, these two elements are in relation, R: yRx means that $\#\{i \in N : \text{for all } z \neq x, xP_iz\} \geq \#\{i \in N : \text{for all } z \neq y, yP_iz\}$; $\#$ means *number of...* These rules ensure the fairness of the most preferable choice and enhance its credibility, for example, by avoiding (or at least disclosing) a *dictatorial* opinion. That is, an expert's judgment should not dominate the others' judgments. Aggregate individual expert judgments differ from judgments made by *juries* at trial. The latter are informed by direct and tangible combinations of theories and facts (evidence), interpretations of those facts under two diametrically opposite point of view (as explained by experts), rebuttals and corrections, and possible conflicting assessments of how these facts should be understood (under a specific application of regulatory or administrative law).[1] To exemplify these assessments, we take a causal chain as consisting of three essential elements on which individual experts either vote and their votes are then summed. These are (see 8.1): {*formal assumptions*}, {*modeling*}, and {*results*}. Individual judgments should at least be better than random (i.e., choices made by flipping an unbiased coin). Each expert judgment on each element of the triplet may be voted as being either *true* (+1), *false* (−1), *undecided* (0). The set of {results} is voted according to: {assumptions} AND {causation} → {results}. Voting on {causation} could account for nonlinearity: +1:=causal association, 0:=mechanistic association, :=−1 statistical correlation. Each judgment (e.g., undecided) may also be weighted (i.e., $\sum_i w_i = 1.0$). Aggregation of the individual experts' judgments obtains through voting rules such as plurality, simple or other forms of majority. {*Modeling*} may imply that {*causation|exposure to XYZ*}. Alternatively, modeling may mean that cause and effect is a nonlinear exposure-response function. A simple majority on this element, {.}, may be that #5 (+1) to #4 (−1) for nonlinearity: five experts vote *True*; the remaining four voted *False*. This set can be decoupled into more specific sub-sets of K&I. For example, suppose that the existing literature and specific ad hoc experiments formalize the micro level causal process as:

$$\{\text{response}|\{\text{toxicant } XYZ|\text{enters blood flow}\} \rightarrow \text{Reactive oxidant}$$
$$\text{species formation} \rightarrow \text{inflammation} \rightarrow \text{coagulation}, \rightarrow \text{impairment} \qquad (8.2)$$
$$\text{of normal vascular function}\}.$$

Unless XYZ is known directly to impair the specific function, the relationship between exposure to XYZ and the biochemical change from XYZ to the by-product that actually causes the impairment should also be accounted. For instance, there may be some evidence that confirms this biochemical change and that the by-product is determined by pharmacokinetic changes from XYZ to XYZ_{Be}. The biologically active dose of the by-product, XYZ_{Be} may be stated as (XYZ_{Be}|PB-PK&XYZ); the vertical line means *given that*. Suppose that a group of nine experts is empaneled to vote on (1), given a specific physical context. What voting rule should be used?

---

[1] The evidence for juries (who are not scientific experts) is introduced under oath; the selection of members of a jury differs from the selection of experts.

And: What is the formal basis for the rule so that the correct rule can be selected? (Mach et al. 2017). As we will discuss in later sections regarding preferences by N individuals between nonoverlapping and uniquely distinct alternatives x, y, and z, the relationships, R, over these should be (at least) pairwise complete (e.g., xRy AND yRx) and transitive (i.e., IF $xRyRx$ THEN xRz). The term xRy is read: *x in relation to y*. In the discussions that follow, the set of choices to be preferentially ordered is $X = \{x, y, z, \ldots\}$. The choice for aggregating individual judgments (for instance, xR_iy) is *a* majority rule: $\#\{i \in N : xR_iy\} \geq \# \{i \in N : yR_ix\}$: the count (#) of preferential relationships between two objects is larger than the count for its alternative.

Choosing how to formally aggregate and combine heterogeneous evidence using (8.2) involves analytical work, as discussed in the previous chapters. The overall process should result in what may be considered to be the *best available evidence* of causation. But, because of incomplete K&I, additional evidence has to be developed to confirm that a catastrophe is not in the *bubbling* phase. For example, under the FDA *marketing order* regulatory process for reduced risk tobacco products, the applicant/producer is an agent-producer for society, who is supervised by this agency. The FDA decides if the producer's data, experiments, and observational studies satisfy the statutory command of being the *best scientific evidence* of reduced risk and consequences and decides to either allow continued production or to cease and desist. This requires that the agent-producer informs – under a strict reporting protocol – the FDA that monitors the course of the scientific K&I over time; the duty on the manufacturer is to immediately report to this agency changes.

8.3.1 Voting

Voting reflects the *weight* assigned to each choice in an admissible set of choices but not the strength of the difference between those choices (which is given by a ranking). In the simplest instance, a set of individual *#yes* and *#no* (within a total count of *#yes* and *#no*) is aggregated over all experts, each of whom has one vote (a more complete discussion about voting is beyond the scope of this chapter). We briefly discuss the *procedural* (List 2012) criteria for majority rule. These criteria are as follows: (i) universal domain, (ii) anonymity, (iii) neutrality, and (iv) responsiveness. The first criterion requires identifying all *logically possible* voting profiles. The second criterion states that permutation of the elements of $V(Vj|A_i)$ should not affect the aggregate judgment, Table 8.1. Third, reversing the votes of a profile reverses the choice. Responsiveness means that, if some votes change, then $V(.)$ should change accordingly. The majority rule satisfies these four axioms (List 2012). Each individual expert ($j \in N$) assigns a vote, $v_{i,j} = (1, 0, -1)$, given her background knowledge on each alternative, A_i ($i > 2$). Aggregation over the N experts is assigned the value 1 when the counts of the 1s (#1) > 0, 0 when the counts equal zero, and −1 when <0. Experts give their judgment; abstention is not allowed, but there can be a minority report. If each choice is judged according to three states (e.g., acceptable, undecided, unacceptable) the number of combinations is $k = 3^N$:

Table 8.1 Voting profiles between alternatives

Alternatives	$V1$	$V2$	$V3$	$...j$	AGGR(#Vj); $j > 2$; simple majority
A_1	...	...	...	...	$\#(V_{j=1}|A_{j=1}) = V_1$
A_2	...	...	$V_i = 2, j = 3$	...	V_2
A_3	...	...	...	...	V_3
$...i$	...	...	...	...	...
		Voters' profiles for i-th-alternative			**AGGR** (Profiles): $V = f(V1, V2...)$

Table 8.2 Choices and judgment profiles

Choices, c_i	Judgment profiles for $N = 2$ individuals		
	$j = 1$	$j = 2$	Aggregation
c_1	1	1	**1**
c_2	1	−1	0
c_3	−1	1	0
c_4	−1	−1	−1

their analysis requires an exponential computational scheme. The analysis, using simple majority over the counts (#) of votes, is depicted in Table 8.1. Shaded areas imply voting according to three criteria (*for, undecided, against*: {1, 0, −1}) and their counts across each voter profiles (1, if # 1 > −1; 0, if # 1 = # − 1; and −1 if (# −1 > −1)). The results from these assessments cannot be generalized to any other circumstance and are said to be *extensional* (rather than *intensional*); each voting profile should be admissible. For example, aggregation by majority rule means that c_1 is selected, Table 8.2.

8.3.2 Ranking

Ordinal ranking of preference over choices is another example of a collective assessment. Ranking is important when choices directly affect public health decisions and the probability of an unknown, but latent and unexpected, catastrophic burden of disease may be present. For instance, assume that a novel product that has been shown significantly to reduce risk and has been granted a marketing order. The evidence permits its production because the probability of an error in the public agency's decision (a regulatory process) is low, but the benefits (reduced risk of disease in population already at risk) is much higher. However, the order also may result, with a low probability, in a DK that may be due to latency of certain chronic diseases in humans. In a precautionary model against the prospective DK, the permit to sell is contingent on the sequential addition of causal K&I over specified time periods. The preventive mode is that the agency can issue a *stop order* at any time.

Table 8.3 Analysis of choices and experts' relationships, AGGR(.) is aggregation

Available choice	Expert 1	Expert 2	Expert 3	$AGG(R_i) = F(R_1, R_2, R_3)$
C_1	xR_iy	xR_iy	xI_iy	xRy
C_2	xR_iy	yR_iz	zR_ix	Violates ordered transitivity
C_3	xR_iy	xP_iy	yP_ix	xPy

Formally to assess choices (more than 2) we use Arrow's five *preference* axioms[2] over choices. These are as follows: (i) universal domain; (ii) ordering (i.e., completeness and transitivity); (iii) independence of irrelevant alternatives; (iv) weak Pareto ordering; and (v) non-dictatorship. Briefly, completeness prevents asymmetric information by requiring all possible profiles to be accounted. *Ordering* implies linear ranking; *irrelevant alternatives* should not affect rational ranking. For *Pareto ordering*, if pairwise ranking yields a choice and that choice is similarly ranked across all pair-wise comparisons, then their aggregate must reflect this. For *non-dictatorship*, aggregate judgments should not be dictated by a single individual expert. Unfortunately, Arrow demonstrated (with two impossibility theorems) that aggregation over three or more choices by more than two individuals cannot0020simultaneously satisfy these five axioms. More importantly, there are no voting methods that do so (i.e., *impossibility* theorems; Arrow, in Dietrich and List 2007a; Dietrich and Mongin 2010). For instance, Condorcet's paradox violates the transitivity axiom because of cycles between preferences in head-to-head or sub-group to sub-group comparisons. These break transitivity. Consider the group $G = A + B + C$ with a strict preference order (P) over three choices, x, y, and z: $xPyPz$. Assume that each sub-group has the same number of individuals and that A ($xPyPz$); B ($yPzPx$), and C ($zPxPy$). The transitivity axiom is violated because the separation into sub-groups results in a preference ranking that is irrational, relative to the aggregate $xPyPz$. Consider the sets of discrete and nonoverlapping alternatives $X = \{x, y, z \dots\}$ and ordered preferential relationships, $R = \{R_1, R_2 \dots\}$ to be *complete* and *transitive*. Suppose that the set of alternatives $\{x, y, $ and $z\}$ leads to finding – for the i-th individual – xR_iy, yR_iz, and so on. Following List (2012), the strength of the ordered relationship may be: *weak* (xR_iy and it can also be that yR_ix) or *strong* (xP_iy & yP_iz); *indifferent* (I) between the two choices, namely, xI_iy and yI_ix. For example, given three choices and three experts, the following situation may arise (i is an individual in N; R, P are aggregations over R_i and P_i), Table 8.3.

In a group of experts, some of them may form an oligarchy. In other words, M experts, $M \subseteq N$, such that $i \in M$; xP_iy implies that xPy and, for some i, xR_iy. It may be helpful to use the *single peakedness* rule in which, given three choices x, y, and z, one of these choices is preferred then either of the other in some descending order from the preferred one on both sides of the *peak*. In some instances, if a complete ordering cannot be achieved, then a partial ordering may be sufficient. This is the case for choice C_2: expert 3 violates transitivity. [3]

---

[2] List (2012) discusses these axioms; unless otherwise stated, we deal with pairwise choices.
[3] Borda's count violates the transitivity axiom.

Interestingly, regarding defining experts, Condorcet's *jury theorem* states that: (i) an expert must have a long-run average probability of being correct that is better than 50–50%; and (ii) experts should be independent. The former quality is called *competence* (Dietrich and List (2004)). Assuming N ($i = 1, 2, \ldots$) voting experts and $x \in \{1, -1\}$ judgments, *competence* means that each vote, v_i, has the property that $pr(v_i = x | X = x) > 0.5$, pr has the same meaning for individuals and judgments, and that the N individuals' votes are probabilistically independent. Condorcet's jury theorem (List 2012) proves that majority voting converges to the correct judgment with probability 1.0, as the number of voters N tends to infinity. Shapley and Grofman (1984) demonstrate optimality (i.e., a *better* choice in terms of a linear ordering of binary choices) for uncorrelated and correlated voting. Consider N (odd) experts; each expert makes a binary choice with pr_i ($i = 1, \ldots, N$) and a linear preference order. If the prior $pr_i = 0.5$, then the weighted voting process, with weight $\log(pr_i/(1 - pr_i)$ maximizes the likelihood that the aggregate judgments of those N experts will converge to the *correct* (more accurate than under other probabilistic alternatives) choice. Following List (2012), Condorcet's independence, if violated, does not invalidate results. However, if expert opinions are highly positively correlated, the reliability of the aggregate results decreases as a function of the magnitude of the correlation (Dietrich and List 2007b, c). Condorcet's *independence* of experts and *competence* definitions can be relevant to understand results generated by expert panels. In other words, each expert's record should be factually established relative to their past judgment on a specific topic. Condorcet's requirements are simple and plausible: they are quantitative rules of thumb rapidly to assess an expert's possible prior stance on an issue. However, particularly for the inevitably correlated K&I informing the judgments of experts, Shapley and Grofman (1984) show that Condorcet's probabilistic competence is not a critical issue if average probabilities and the normal distribution are used. This finding is useful for assessing the impact of routine events. For non-routine events, where the distribution functions may only have a right and thick tail, cannot be assumed to be asymptotic, more complicated criteria may have to be used.

Example We take (\{assumptions\} AND \{causation\} $\rightarrow$ \{results\}); assume five individuals with corresponding set of Condorcet competences \{0.9, 0.9, 0.6, 0.6, 0.6\}; and three independent aggregate judgments. If the largest competence determines the aggregate choice, then 0.9 is selected. This result is close to being dictatorial, but for the fact that two experts have the same judgment. Simple majority rule yields 0.6. Ranking using the log(odds) weighted probabilities yields 0.90; 0.88, and 0.93.

8.3.3   Discussion

A critical question is: What regulatory probabilistic standard of acceptance should the scientific evidence overcome? The question is about the balancing of the evidence for or against a causal assertion by assessing it relative to a probabilistic acceptance threshold, for instance *more likely than not*. This form of evidence balancing, discussed in Chap. 2, assesses whether the evidence satisfies a civil law the burden of proof. However, an entity may require the producer of the risk to demonstrate that some level of mandatory mitigation *significantly* reduces risk. The term *significantly* implies a larger probability threshold that the *more likely than not* standard of proof. Moreover, there no middle ground, the rule is binary: either the required burden is met, or it is not. It may often be preferable to account for a third state of the evidence: one that comports having to wait for additional evidence before acting. The reason is that scientific evidence is continually evolving and thus a binary choice is inconsistent with scientific reasoning, particularly for prospective events and their consequences. There is a difference between computed probabilities and probabilities used as thresholds in regulatory law. The latter is directed by legal principles; the former is the result of the algebra of probabilistic events and is computed in an analysis. Probabilities weigh the quality and the strength of evidence, not its quantity. In legal contexts, the acceptability of the totality of the evidence may be stated as being: (i) *more likely than not*, (ii) *clear and convincing*, or, (iii) *beyond a reasonable doubt*. The weight of the evidence that overcomes binary alternative runs from >0.50 to ≤1.0. Where the numerical values actually fall is not stated; attempts to do so have proven that the fact finder (either a judge or a lay jury) violates the precepts of probability theory, we have discussed these in Chap. 2. Policy could plausibly assert thresholds such as (i):=0.51, (ii):=0.70; and (iii):=0.90 depending on the context and type of catastrophic event that the policy is directed against. Because the balancing that we are concerned with is probabilistic (and inherently Bayesian), an intermediate state – the *do-nothing* option – may be needed for sound decision-making. An initial step toward accounting for this state is to set upper and lower uncertainty limits on each threshold value, rather than use a single probability value for each. Numerical standards are routinely developed from assessing results of merit from integrated scientific research (e.g., they are developed by the agency for rulemaking and may be relatively small). In other words, an agency selects specific evidence from the literature, assesses it, may itself conduct or otherwise require additional research, develops a damage function (e.g., a dose-response model for diseases), develops safety factors for engineered structures, or develops a weighted exposure limit for occupational health. These support the agency's aggregate policy judgment that combines causation with probabilistic analysis. In many novel situations, the evidence is much less credible that it is for established situations. We turn to this issue next.

8.4   Aggregating Vague and Contradictory Evidence: A Bayesian Method

The issue that we address concerns the representation of complex, qualitatively *soft* and vague (e.g., jointly contradictory, ambiguous, and personal) K&I. The objective is to provide a replicable and sound method to represent this information and combine (i.e., to aggregate and fuse) it through some scientific method. An important aspect of the methods we discuss is that they help to deal with situations where K&I may be akin to hearsay or where the source of the information is incapable to provide crisp characterizations of K&I. The form of aggregation we discuss next ranges from the purely subjective, for example an expert may conclude that *I believe that A and B occurred*, without adding more, to quasi-Bayesian methods (Davis et al. 2016). The focus of the discussion is on assessing the relevance of the K&I available and making it useful to the stakeholders, in other words, stories and anecdotes. This less precise, but realistic state of K&I, is critical to the more complete understanding of many of the aspects of prospective catastrophic incidents by combining with more formal K&I. The method we discuss deals with evenly spaced values, integers measured on a scale from 0 to 10. The fusion is quasi-Bayesian (uses Bayes' theorem and probabilities that can be called *credences*, measured in the closed interval [0, 1]), which we still symbolize as pr; pr(X|Y) means that X is *credentially* probable given Y. This formalism preserves the structure of the qualitative information by maintaining their salience, quality, and credibility (Davis et al. 2016). We consider two situations. In the first, a *report* by an author may contain multiple estimates of the effect of one or more risk factors on the consequences. In the second, several reports by one or more authors provide (presumably independent) more complete estimates. In the discussions that follow, a single risk factor is identified by the letter X; if there are several risk factors, the subscript i identifies them (i.e., i = 1, 2, ..., k) such that {X_1 ... X_k}. The structure of the quasi-Bayesian analysis consists of priors and likelihoods. Priors, the *credence* that a factor X_i is valued as a number between 0 and 10 (integers) and the likelihoods, conditional *credences*, are shown in Tables 8.4 and 8.5.

Table 8.5 depicts the likelihoods, which sum to 1.0 across each row and each column, developed by an expert assessing a single factor, for example, factor X_i. The numbers in the example that follows are provided by a single expert; multiple experts provide their own table of estimates of these or other likelihoods, fully elaborated by Davis et al. (2016).

This framework provides the background for assessing K&I about the consequences from exposure to a set of risk factors as found from either a single or in multiple reports. These reports may range from oral descriptions to official transcripts of questions. The aggregate probabilistic uncertainty can be represented by likelihood tables in which the likelihoods indicate symmetry or asymmetry of opinions, as discussed in Davis et al. (2016). The probability scale [0, 1] maps to credence levels [[0, 10, integers]. Davis et al., map from levels [1, 2, 3, 4, 5] to reported probabilities. The sum of each row probabilities equals 1.0.

Table 8.4 Prior probabilities, pr(X_i), and hypothetical values, 0.60, from a single report

Credence level [0, 10]→Factors ↓	0	...	10	Total probability
X_1	pr = 0.60	...	...	1.0
...		...	...	1.0
X_i		...	pr(X_i)	...	1.0
...			...	...	1.0
X_n		...	...	...	1.0

We only show a single credence value and indicate that, for any factor, the sum of the credences adds to 1.0

Table 8.5 Likelihood values, pr(X_i|E), hypothetical values, 0.30; E is the new evidence conditioned on the old evidence, assuming two reports (the baseline and the new reports), dealing with the same risk factors

Likelihoods (conditional probabilities)	New report			Total row probability
Baseline report	0	...	10	NA
0	...	...	...	1.0
...	...	pr(E|X_i)	...	1.0
10	...	...	pr = 0.30	1.0
NA	1.0	1.0	1.0	1.0

The column numbers and the row number must each add up to 1.0. The higher the probability number in a cell the higher the confidence in the matching between the two reports
NA is not applicable

8.4.1   Comment

We can form weighted ($0 \leq w_i \leq 1$) linear combinations of values for the consequences, C. That is, $C = \sum_i w_i X_i$ where the consequence is valued on an integer scale [0, 10]. The formula can be further modified to account for probabilistic uncertainty using subjective probabilities (pr) as well as weights: the combining formula yields $C* = \sum_i w_i \text{pr}_i X_i$. A threshold, E_F, can also be added, further modifying the consequence: $C** = g(F, e_f)(\sum_i w_i \text{pr}_i X_i)$, in which e_F is a threshold that marks the first value of the function F. Specifically, $g(F, e_f) = 0$, if it is the minimum of $(F - e_f)$; otherwise it equals 1. Thresholds can be determined by policy and past assessments of K&I. A threshold can be a specific value that identifies a point ($d* < d$) where $d*$ is the boundary that separates what is known not to contribute to the consequence to what is known to begin the effect. Davis et al. (2016) also suggest that it may be relevant to account for primary risk factors, P, and secondary ones. The separation of primary from secondary risk factors can be either subjective or objective. That is, linear weight sums, LWS, threshold LWS, TWLS, and Primary Factors (PF) analyses can be used to fuse vaguely reported K&I. Specifically, for an outcome O, the threshold LWS, TWLS, is calculated as $O = Q(F, Th)*\sum_k w_k F_k$, in which Th is a threshold either equal to 0 or different from 0 (Davis et al. 2016). The evidence is gathered from several independent *reports*, each report is indexed by the subscript I (Davis et al. 2016):

$$\text{pr}\left(X_i|E\right) = \frac{\text{pr}\left(E|X_i\right)\text{pr}\left(X_i\right)}{\sum_{n=1}^{n}\text{pr}\left(E|X_n\right)\text{pr}_n\left(X\right)} \tag{8.3}$$

In this equation, X_i is a factor the consequences of which are measured on the equally spaced [0, 10] scale, taking integer values on this axis from low to high values, for example, 1, 3, 5, 7, 9. Division by zero is not permitted. Note that $\text{pr}(C) = \sum_i w_i(\text{pr}_i X_i)$ or 0, if $X_i < e_F$. As we have discussed, aggregation using Bayes' theorem, particularly when there are conflicting, vague, or inaccurate findings, complicates probabilistic analysis. Moreover, subjective evaluations may be affected by issues such as: (i) values are expressed using crisp numbers without discussion, (ii) cognitive biases affect the correct understanding of probabilities, and (iii) correlations between the attributes may not be known to the analysts. Davis et al. (2016) provides the details about how to deal with these and other issues and suggests an important way to assess aggregate judgments using the maximum entropy, minimum penalty, MEMP method. Most briefly, the objective is to maximize the entropy of information of a set of discrete quantities of information, characterized by probabilistic uncertainty, while minimizing overfitting to the data. For example, individual A is concerned with the set of consequences $\{C_j\}$: $\text{pr}(C_A) = \{\text{pr}(C_1), \ldots \text{pr}(C_j), \ldots \text{pr}(C_m)\}$. The values of C_j may be generated by independent sources. The measurements on $\{C\}$ are made on the integer, equally spaced, scale $[0, 10]$; equally spaced intervals can be represented by single values: for example, $[0, 2] := 1$. Entropy of information is an average representation, for the ith expert in a group of experts A, $\ldots G, \ldots Z$, namely[4]:

$$H_i(C) = \sum_{j=1}^{m} \text{pr}(C_j) \log_2 \left(\frac{1}{\text{pr}}(C_j) \right) \tag{8.4}$$

The number of risk factors is indexed by j. $H(C)$ is measured in "bits" of information when the logarithms are to the base 2. We use its more common expression:

$$H_i(C) = -\sum_{j=1}^{m} \text{pr}(C_j) \log_2 \left(\text{pr}(C_j) \right) \tag{8.5}$$

Example Suppose that there are two outcomes. Researcher A reports equal probabilities (0.5), each outcome is valued 1; her $H = 0.5*\log_2(0.5)*(1) + 0.5*\log_2(0.5)*(1) = 1.00$. B reports his H as: $0.75*\log_2(0.75)*(1) + 0.25*\log_2(0.25)*(1) = 0.81$. Thus, entropy of information reaches a maximum when there is less information: the discrete uniform distribution (0.5, 0.5) is less informative because it has a larger entropy than its alternative with mass distribution (0.75, 0.25). Entropy of information is additive, $H(Z) = H(X) + H(Y)$, provided that X and Y are stochastically independent and that $H(0) = 0$. In the practical context of assessing alternative reports, the problem can be stated as seeking the set of probabilities that maximizes entropy and accounts for probabilistic uncertainty. In other words, some probabilistic outcomes or judgments can be different from being uniformly distributed, others some may be so distributed. The issue is how to solve for all probabilities in the set. For n values, the upper bound is $H(C) = \log_2(n)H(C)$. MEMP can account for conflicting reports. For instance, A reports the mass distribution function (0.75, 0.25) but B reports (0.25, 0.75): a clear conflict, particularly when viewed against (0.5, 0.5). The quality and reliability of the reports combine as: quality:= (credibility [0, 1]*salience [0, 1]).

---

[4] By assumption: $0*\log_2 0 := 0$ to avoid computational difficulties (e.g., indeterminacy).

8.5 Conclusion

We discuss the basis for aggregating individual expert opinions through voting or ranking, introduced in Chap. 7. We discuss both ranking and voting methods as used in the aggregation of scientific judgments and briefly discuss some of their main characteristics in the context of limited K&I. We find that limited K&I can be useful, and thus merits some discussion, particularly when the stakes are high and little else may be available. We then summarize alternatives to represent aggregated K&I in the context of the sequence inputs, assumptions, methods, and results. Uncertainty about the elements of this sequence may be weighted *credences*, rather than probability measures, because the event space is not known. The reason for not knowing the event space is it is not reported or, if reported, it may be biased. Credences are quantitative degrees of belief in propositions provided by agents or other individuals familiar with the subject matter at hand. These measures can be weighted to indicate their reliability and thus form a plausible way to assess what may turn out to be the only evidence available regarding the prospective impact of complex cascading catastrophic incidents. These aspects are particularly important when there are conflicts between experts, causation is uncertain because alternative sources of K&I conflict with one another, and the inputs and their sources inputs are questionable. When this type of weak K&I is available, and little else is known, other than first principles and some regularities, it is informative even though it does not reach the level associated with the *best available evidence*. The formalism we discuss should enhance science-policy discussions by adding to the means for assessing evidence in a common probabilistic frame of reference. Recall that probabilities are the most commonly used as measures of uncertainty in regulatory law and public decision-making. A reasonable step toward assessing uncertain causation, which might otherwise remain couched in vague terms, is to enlarge the scope of probabilistic analysis to be consistent with credences, degrees of belief, upper and lower probabilities (DS measures), frequencies, and their common analytical basis.

References

P.K. Davis, W.L. Perry, J.S. Hollywood, D. Manheim, *Uncertainty-sensitive heterogeneous information fusion: assessing threat with soft, uncertain, and conflicting evidence* (RAND Corp., Santa Monica, 2016), 143

F. Dietrich, C. List, Arrow's theorem in judgment aggregation. Soc. Choice Welf. **29**, 19–33 (2007a)

F. Dietrich, C. List, Strategy-proof judgment aggregation. Econ. Philos. **23**, 269–300 (2007b)

F. Dietrich, C. List, Opinion pooling on general agendas (2007c) [available online (pdf)]

F. Dietrich, P. Mongin, The premise-based approach to judgment aggregation. J. Econ. Theory **145**, 562–582 (2010)

T. Dietz, Bringing values and deliberation to science communication. PNAS **110**, 14081–14087 (2013)

S.M. Hsiang, K.C. Meng, Reconciling disagreement over climate–conflict results in Africa. Proc. Natl. Acad. Sci. U. S. A. **111**, 2100–2103 (2014)

C. List, The theory of judgment aggregation: an introductory review. Synthese **187**(1), 179–207 (2012). ISSN 0039-7857). https://doi.org/10.1007/s11229-011-0025-3

K.J. Mach, M.D. Moscandrea, P.T. Freeman, C.B. Field, Unleashing expert judgment in assessment. Glob. Environ. Change **44**, 1–14 (2017)

M.G. Morgan, Use (and abuse) of expert elicitation in support of decision making for public policy. Proc. Natl. Acad. Sci. U. S. A. **111**, 7176–7184 (2014)

L.S. Shapley, B. Grofman, Optimizing group judgmental accuracy in the presence of interdependencies. Public Choice **43**, 329–343 (1984)

S.G. Straus, A.M. Parker, J.B. Bruces, J.W. Dembosky, The group matters: a review of the effects of group interaction on processes and outcomes in analytic teams. Working Paper No. WR-580-USG (RAND, 2009), https://www.rand.org/content/dam/rand/pubs/working_papers/2009/RAND_WR580.pdf

The White House, Executive Order, EO, (83 FR 18768), Proposed Rule, Data Transparency (2018)

D. Von Winterfeldt, Bridging the gap between science and decision making. PNAS **110**, 1055–14066 (2013)

G. Wong-Parod, T. Krishnamurti, A. Davis, A. Schwartz, B. Fischoff, A decision-science approach for integrating social science approach for integrating climate and energy solutions. Nat. Clim. Chang. **6**, 563 (2012). https://doi.org/10.1038/CLIMATE2912

Index

© Springer Nature Switzerland AG 2020
P. F. Ricci, *Analysis of Catastrophes and Their Public Health Consequences*,
https://doi.org/10.1007/978-3-030-48066-0

Printed in the United States
by Baker & Taylor Publisher Services